ORIGINS OF INDUSTRIAL ENGINEERING

LIST OF MAJOR DONORS

The Institute of Industrial Engineers thanks the following people whose generous support helped make this book possible.

W. E. Alberts
Irwin E. Alperin
Ray L. Anderson
Michael J. Armenia
Henry Baird
Larry J. Baker
David A. Balasa
Baltimore Gas and Electric
James W. Barany
Robert L. Bedwell
David L. Belden, Ph.D.
Heinz G. Bieberstein
Dr. and Mrs. John E. Biegel
George W. Billmyer, III
A. Donald Bird
Leland T. Blank, Ph.D., P.E.
B. B. Bleckley, III
James E. Bodi, P.E.
Joel Borden, P.E.
Robert W. Boscher
William H. Brown
Philip Brumbaugh
Ellsworth S. Cabot
Earl D. Cantrell
V. Sam Carlile
Gordon B. Carson
Tod Carter, P.E.
Kenneth E. Case, Ph.D., P.E.
James A. Chisman, Ph.D.
Charles E. Clark, Jr., P.E.
Franklin H. Cloud
J. Logan
James M. Cogdill
The Cooper Group
George M. Copeland, P.E.
George L. Crowder
Adam W. Cywar
Robert P. Davis
Santo L. DiVita
Archie D. Dixon
John F. Douglas, P.E.
Dan C. Doulet
Donald O. Eisele
J. T. Elrod
Salah E. Elmaghraby, Ph.D.
H. O. Enders
Jack W. Enterline, P.E.
Wolter J. Fabrycky, Ph.D., P.E.
Harris A. Fein
Floyd E. Fields
Gerald A. Fleischer, Ph.D., P.E.
H. J. Foehringer
John R. Freeman
John D. Frisk
R. William Gardner
Sidney Gilbreath, Ph.D.
Leroy O. Gillette, P.E.
Everett A. Greenli
Mikell P. Groover, Ph.D., P.E.
Robert P. Groseclose
Frank M. Gryna
Bernard H. Guthrie
William E. Hammer, Jr.
V. L. Hastings
Dale B. Hartman
Frederick L. Haynes, P.E.
Samah Helal
Patrick L. Henry
L. Jay Hinton
William A. Hoikala
C. Red Hollander
Jerome M. Holzman, P.E., P.P.
John R. Huffman, P.E.
Richard C. Hutchinson
James R. Innis
Roland F. Jennings
David W. Johnson

Iver C. Johnson, P.E.
George E. Kane
William J. Kennedy, Jr., Ph.D., P.E.
Tarek M. Khalil, Ph.D., P.E.
M. Glenn Klutz
Ong-Arj Kriengkripetch, P.E.
Harvey R. Kulin
Sidney N. Lakin
N. A. Lamberti, Ph.D.
Henry M. Landes, P.E.
Richard A. Leshuk
Soter G. Liberty, P.E.
Kjell Liljeblad
Harold R. Lindemann
Steve C. Lo
Hall A. Logan
Joseph J. Logan
Howard L. Loveless
Jesse Lee Martin, P.E.
William L. Maxwell, Ph.D.
F. Wayne McKenzie
William R. McReynolds
Wilbur L. Meier, Jr., Ph.D., P.E.
Gerald Mendenhall, P.E.
Charles M. Merrick
Joseph P. Milillo
Joe H. Mize, Ph.D., P.E.
Harold S. Mohler
L. L. Mullins
Gerald Nadler, Ph.D., P.E.
Seetharma L. Narasimhan, P.E.
E. T. Newbrough
Albert R. Noel, Jr.
Keytack H. Oh
Flora and Robert O'Hagan
Alfred Opengart, P.E.
Irvin Otis, P.E.
Murl Wayne Parker, P.E.
Newton Parks, P.E.
George M. Parks, Ph.D.
Russell C. Peters, P.E.
C. Winfred Powell
Harold F. Puff
Joseph H. Quick
Richard E. Reeder
William A. Reed
Robert E. Reiss
Jack B. ReVelle, Ph.D.
John R. Rogers Associates
Vinod K. Sahney, Ph.D.
Martin R. Sanders, P.E.
Neal F. Schmeidler
Harry A. Seifert
Sol Seltzer, P.E.
F. Stan Settles, Ph.D.
William H. Sherman
J. Landon Short
Gene R. Simons, Ph.D.
William D. Shadle
Clifford E. Smith
John N. Snider
Richard J. Steele, CMC
Randall J. Steger
Christian W. Stengel, Jr.
Arthur E. Stukey
William Swart
Stewart Fraser Taylor
Technical University of Nova Scotia
Camiel Thorrez
Gerald Thuesen, Ph.D.
James A. Tompkins, Ph.D.
University of Pittsburgh, P.E.
Ed Utterback
Joseph Voldrich
Paul M. Wallack, Ph.D., P.E.
Lisa Weil, P.E.
Ralph A. Welch
Vance K. Wilkinson
A. William Wortham
William Wrennall
Andrew C. Zunzer

ORIGINS OF INDUSTRIAL ENGINEERING

The Early Years of a Profession

Howard P. Emerson &
Douglas C.E. Naehring

Published by
Industrial Engineering & Management Press
Institute of Industrial Engineers®

40 Years of Service to the Industrial Engineering Profession
1948 · 1988

25 Technology Park/Atlanta, Norcross, Georgia 30092
(404) 449-0460

Printed in the United States of America.
92 91 90 89 88 5 4 3 2 1

Library of Congress Cataloging-in-Publication Data

Emerson, Howard P., 1901-
Origins of industrial engineering.

Bibliography: p.
Includes index.
1. Industrial engineering--History, I. Naehring, Douglas C. E., 1924- II. Title.
T55.6.E44 1988 670'.9 88-27392
ISBN 0-89806-097-4

Table of Contents

Preface

It is important for a profession to respect its past and those who contributed to it. The knowledge of early objectives and subsequent developments can be of invaluable assistance in future growth, hence, a recording of the events of the early days of industrial engineering as a profession is in order.

The evolution of a profession requires support from many areas. There are individuals who originate the concepts and objectives; there are others who apply them. Educational programs and curricula need to be developed in universities to educate those who will carry on the profession. For a profession such as industrial engineering to mature, private industry must experiment with applications of the new concepts, aided by consultants who provide training and advisory assistance. Government recognition and applications of the profession in its shops are also essential.

A society or professional organization must be established to provide a community for those who identify with the profession and to establish a forum for the varied concepts, practices, and philosophies. A major contribution of such a society is its publications, which must be an authoritative source of information about accomplishments within the profession. All of these things have come to pass in the growth of industrial engineering.

This book is an account of each of these steps in the founding and growth of industrial engineering. Emphasis is on the period of the profession's earliest developments through the founding, in 1948, of the American Institute of Industrial Engineers (now the Institute of Industrial Engineers). Since the Institute's founding, much of the profession's history and many of its records have been kept and noted in Institute publications. A formal definition of industrial engineering was accepted by the Institute's Board of Trustees in 1955 and was revised in 1985:

> Industrial engineering is concerned with the design, improvement, and installation of integrated systems of people, material, information, equipment, and energy. It

> draws upon specialized knowledge and skills in the mathematical, physical and social sciences together with the principles and methods of engineering analysis and design to specify, predict and evaluate the results to be obtained from such systems.

This book is published to commemorate the 40th anniversary of the founding of the Institute of Industrial Engineers. It is not intended to be a comprehensive history, for it would be difficult to improve on publications such as Frank B. Copley's two-volume work, *Frederick W. Taylor* (1923 Harper and Brothers, New York).

An understanding of the profession as it exists today can be gained by knowledge of the early events and significant individuals and organizations involved in the founding of industrial engineering. This book includes important records that are no longer easy to find as well as references to additional sources. Also included are stories of how the early pioneers in industrial engineering and management formed this new field and the concepts and techniques they developed. This book is not only for practicing industrial engineers, but also for industrial engineering students, those who work with industrial engineers, and those who have retired from the profession. It is intended to ensure that the legacy of the early years of the profession will be preserved.

Howard P. Emerson
Douglas C. E. Naehring

1. A Profession Is Born

In 1912, the originators of industrial engineering and the early pioneers in its development; the first educators to present courses, write texts, and develop curricula; consultants who introduced industrial engineering and scientific management to industry; and the managers and representatives of the first industries to adopt the new concepts had gathered at the annual meeting of the American Society of Mechanical Engineers (ASME). This one meeting, held in New York City, provided more information about the origin of industrial engineering than any other event.

The all-day session of Friday, December 6, 1912, began with a presentation of the Majority Report of the Subcommittee on Administration titled, "The Present State of the Art of Industrial Management." Although this was the last day of the three-day annual meeting, attendance was reported to be large. This report and the subsequent discussion provide both insight and understanding about the relative contributions of the individuals involved in a unique new profession: industrial engineering.

Present were the profession's originators and pioneers: Frederick W. Taylor, Frank Gilbreth, Henry Gantt, and Henry Towne. The profession's first educators included Hugo Diemer, who started the first continuing curriculum in industrial engineering at Pennsylvania State College in 1908, and William Kent , who organized an industrial engineering curriculum at Syracuse University in the same year. Others in attendance were Dexter Kimball, who presented an academic course in works administration at Cornell University in 1904, and C. Bertrand Thompson, an instructor in industrial organization at Harvard, where the teaching of Taylor's concepts had been implemented.

Consultants at the meeting included Carl Barth (who recommended that Taylor's concepts be taught at Harvard), S. E. Thompson, and H. K. Hathaway, all of whom taught or lectured at Harvard. Industrial managers included John Aldrich of the New England Butt Company, who presented the first public statement and films about micromotion study; James Dodge, chairman of the committee and president of the Link-Belt Company; and Henry Kendall, who spoke of experiments in organizing personnel functions as part of scientific management in industry.

Two editors present were Charles Going of *The Engineering Magazine*, and Robert Kent, editor of the first magazine with the title of *Industrial Engineering*. Lillian Gilbreth, wife of Frank Gilbreth, was perhaps the only pioneer absent. At that time, women were not admitted to ASME meetings.

The 1912 conference, the committee's scholarly presentation, and the discussion by various individuals whose comments showed their respect and appraisal of the individuals' contributions, answer many questions about the origins of industrial engineering.

How Did Industrial Engineering Start?

Recognition of Taylor as "the father of scientific management" has been challenged at times. For example, in the *Golden Book of Management*, edited by Lyndall Urwick (who agrees with the title that has been bestowed on Taylor), statements by Charles Babbage are compared with those of Taylor. The comment is made that in many respects Babbage "...anticipated some of the most important findings of F. W. Taylor in the United States, though Taylor was quite unacquainted with his work" (Urwick 1956). However, a distinction between the works of Babbage and Taylor cannot be made on the basis that Taylor did not know of Babbage's work. Babbage's book *On the Economy of Machinery and Manufactures* (1832) had been published in Taylor's home town of Philadelphia, and the committee presenting the report at the 1912 annual meeting of the ASME outlined some of Babbage's conclusions. In fact, the appendix to the report included a table by Babbage which Taylor discussed at the meeting.

Table 1, a reproduction of Babbage's table, summarizes the seven operations in the manufacture of straight pins. In addition to cost data, the table included the time to produce one pound of pins for each of the operations and the total time of the entire process. Through this table, Babbage extended the work of Adam Smith on the division of labor by showing that money could be saved by using women and children for the less skilled operations instead of skilled, and accordingly, higher paid men to perform all seven operations.

However, Babbage, who visited factories mainly as an observer, did nothing to try to improve the methods of making the product, to reduce the times required, or to set standards of what the times should be. This was to be the original contribution of Taylor and is in essence the beginning of industrial engineering. Taylor explained this at the ASME meeting.

> It is true that the form of Tables 1 and 2, (note: "Table 2" is reproduced as table 1 herein) given in the Appendix to the

Committee's report is similar to that of the blanks recording time study, but herein the resemblance ceases. Each line in Table 2, for instance, gives statistics regarding the average of the entire work of an operative who works day in and day out, in the running of a machine engaged in the manufacture of pins. This table involves no study whatever of the movements of a man, nor of the time in which his movements should have been made. Mere statistics as to the time which a man takes to do a given piece of work do not constitute "time study". "Time study", as its name implies, involves a careful study of the time in which work ought to be done. In but very few cases is it the time in which work actually was done.

Previous to the development of "time study" in the Midvale Steel Works, there have in all probability been many instances in which men have carefully studied and analyzed the movements of other men, and have timed them with watches. (No such instances have, however, come to my personal attention.) Any such former work was without doubt confined to isolated cases, and was of short duration; and (most important from the historical point of view) it did not lead to the development of a new trade, or, more properly, to a new scientific occupation, "the profession of time study" ("The Present State of the Art of Industrial Management" 1912).

Was Leonardo da Vinci the First Industrial Engineer?

This question was raised in the January 1969 issue of *Industrial Engineering* in an article by Mario Consiglio. He quoted from a manuscript discovered in a European museum showing that Leonardo da Vinci investigated the art of shoveling dirt. Taylor's experiments on shoveling are often considered to be the classic example of applying science to a commonplace activity that has been performed by workers since earliest times, and to be the beginning of the industrial engineering approach. Did da Vinci anticipate this? Consiglio wrote that a good worker, according to da Vinci, could transfer five hundred shovels of loose dirt per hour. If a common shovel of dirt weighed ten pounds, a man could move five thousand pounds of dirt per hour.

However, Taylor's comments about Babbage's table also apply here. Da Vinci did no more than observe a worker and then estimate using a timepiece and a measuring scale the amount of dirt moved in a given

TABLE 1.

Manufacture of Pins
English Manufacture

Pins, "*Elevens*," 5,546 weigh one pound.
Pins, "*One dozen*," 6,932 weigh twenty ounces and require six ounces of paper.

NAME OF THE PROCESS	Workmen	Time for making 1 lb. of Pins	Cost of making 1 lb. of Pins	Workman earns per day s.	Workman earns per day d.	Price of making each part of a single Pin, in Millionths of a penny
		Hours.	*Pence.*	*s.*	*d.*	
1. Drawing Wire .... (§ 224.)	Man. .	.3636	1.2500	3	3	225
2. Straightening wire ....(§ 225.) {	Woman	.3000	.2840	1	0	51
	Girl. .	.3000	.1420	0	6	26
3. Pointing................................. (§ 226.)	Man. .	.3000	1.7750	5	3	319
4. Twisting and cutting Heads.. (§ 227.) {	Boy. .	.0400	.0147	0	4½	3
	Man. .	.0400	.2103	5	4½	38
5. Heading ....(§ 228.)	Woman	4.0000	5.0000	1	3	901
6. Tinning or Whitening .... (§ 229.) {	Man. .	.1071	.6666	6	0	121
	Woman	.1071	.3333	3	0	60
7. Papering .... (§ 230.)	Woman	2.1314	3.1973	1	6	576
		7.6892	12.8732			2320

Number of persons employed: Men, 4; Women, 4; Children, 2. Total, 10.

Source: Charles Babbage, *On the Economy of Machinery and Manufactures.* Reprint of Fourth Edition originally published in 1835. (Augustus M. Kelley, Bookseller, New York, 1963) p. 184.

time, presumably so that he could determine how many men to assign to earth-moving projects. He accepted the methods and the time of the ordinary worker without attempting to improve the job, eliminate wasted motions, or set a scientific, engineered time standard. Thus, neither da Vinci nor Babbage can be considered to have started industrial engineering.

Was Frederick Taylor the Father of Industrial Engineering?

In his discussion at the 1912 ASME meeting, Taylor stated that "Time study was begun in a machine shop of the Midvale Steel Company in 1881" ("The Present State of the Art of Industrial Management" 1912). Taylor had started as a foreman in the shops of the Midvale Steel Company in 1880 and had obtained permission from the president, William Sellers, "to make a series of experiments to investigate the laws of cutting metals" (ASME 1907). He expected that this effort would take six months. Instead, the work continued for twenty years, not only at the Midvale Steel Company but in other companies. The findings were published in Taylor's ASME presidential address, "On the Art of Cutting Metals," in December 1906 (*Transactions of the American Society of Mechanical Engineers* 1907). A second volume had to be added to the *Transactions* that year to incorporate this first scholarly study of applying science to a representative industrial activity. Taylor thought of this massive investigation, however, as just another element in the broader subject of shop or industrial management.

What Were Frank and Lillian Gilbreth's Contributions?

In his discussion of the committee's 1912 report John Aldrich, manager of the New England Butt Company of Providence, Rhode Island, stated, "It is impossible to make up standards without first making a careful study of the correct motions and of the times required to make them." He added that his

> ...company was now using a method of time and motion study which has not been used before, and which not only gives more accurate results than have been heretofore possible, but also enables us to discover methods that are much more economical. This method consists of taking pictures of the various operations with a special motion picture camera, and photographing in each picture a clock of special design showing minute divisions of time ("The Present State of the Art of Industrial Management" 1912).

He displayed pictures of the machine, the special clock (now called a microchronometer or wink counter), and sections of films that enabled an analyst to improve a job. This new approach to work design had been originated by Gilbreth, who explained the technique at the December meeting.

Since working on his first job in 1885, Gilbreth had noticed that each worker performed work using his own methods, although it was possible that one method might be better than others. His first experiments were made on the motions used in bricklaying which he conducted while in the contracting business. The results of these experiments were published beginning in 1908. Gilbreth continued experimental work as a management engineer at the New England Butt Company over a period of five years (1912-1917). There, he established a system of "scientific management as a whole" and carried out his research in motion study.

In his discussion at the 1912 ASME meeting, Gilbreth said,

> As brought out in the report, the importance of transference of skill was realized many years ago. Studies in division of work and in elapsed time of doing work were made by Adam Smith, Charles Babbage, M. Coulomb, and others, but accurate measurement in management became possible when Mr. Taylor devised his method of observing and recording elementary unit net times for performance with measured allowance for fatigue.
>
> It is now possible to capture, record, and transfer not only skill and experience of the best worker, but also the most desirable elements in the methods of all workers ("The Present State of the Art of Industrial Management" 1912).

Gilbreth's later division of manual motions into "therbligs" led not only to micromotion study but, as both he and Taylor predicted, to the development of modern-day predetermined time systems. His process flow chart, with recommendations for analyzing each operation, transportation, storage, and inspection step, gave an overview of the whole production system. It provided the means to improve a system, including the material handling, inventory control, and inspection procedures, through operations analysis. Thus Gilbreth extended and put on a sounder basis techniques for designing and standardizing methods for work--the foundation of industrial engineering.

Frank (posthumously) and Lillian Gilbreth were later awarded the Gantt Medal of 1944, a joint award from the ASME and the American Management Association "in recognition of their pioneer work in management, their development of the principles and techniques of motion study, their application of those techniques in industry, agriculture and the home, and their work in spreading that knowledge through courses of training and classes at universities" (*Mechanical Engineering* 1944).

Was There Any Controversy?

Skeptics of the new practices for improving industrial systems were also in attendance at the 1912 ASME meeting. Barth, a Taylor follower who had applied scientific management at the Link-Belt Company under Taylor's guidance, indicated (referring to the use of motion picture equipment) that he was still "open to conviction regarding the use of any machinery of this kind in connection with time study" ("The Present State of the Art of Industrial Management" 1912).

Kent, who had published Gilbreth's book on motion study in serial form prior to its publication in 1911, retorted that "micromotion study is the most powerful tool ever offered the engineer to measure efficiencies of the worker" ("The Present State of the Art of Industrial Management" 1912).

These statements may be what the committee called "...an unfortunate spirit of intolerance..." ("The Present State of the Art of Industrial Management" 1912) between management experts. The committee felt that the same spirit of mutual helpfulness should prevail in the management field as it did in other engineering fields. Ralph Barnes, in the sixth edition of his book *Motion and Time Study*, states that, "the controversy over the value of these two techniques, time study and motion study, has largely passed; they are inseparable" (1968). Motion and time study has now broadened into the concepts of design and measurement of work, which is the subtitle of Barnes' later editions of his text. His books give full credit to Taylor and Gilbreth as the originators of this important phase of industrial engineering.

When Did Industrial Engineering Start?

If there is a time or date for the conception of this new profession, and if industrial engineering is considered to be the application of science to the design of work and production systems, then the decade of the 1880s can be designated as the starting point.

In 1881, modern time and motion study started under Taylor's guidance. In 1885, Gilbreth began to develop the first rules and techniques of methods improvement. Each in his own way provided original, although complimentary, ideas on the design of work and the measurement of human productivity based on the use of machines and analytical study techniques, thereby laying the foundation for the industrial engineering profession. From Taylor's experiment on shoveling and Gilbreth's study of bricklaying (described in detail in Chapter 2) grew the need for departments of industrial engineering to devise methods of work and set standards for output, quality, and cost; for personnel departments to identify and describe jobs and to select and train workers for them; and for production control departments to schedule work flow using engineered standards.

By researching and conducting experiments in work methods and measurement, a whole range of management responsibilities was opened up. No new ideas of such power or scope as those of Taylor and the Gilbreths invaded the profession until the "quiet revolution" following the Second World War that brought in contributions from the disciplines of statistics, operations research, and computer applications (Schultz 1970). Still, the basic data of work measurement was needed if the mathematical models and computer calculations were to have any practical meaning or decision-making power.

The analytical approach to management problems conceived in the 1880s by Taylor and the Gilbreths gave birth to the new engineering profession. The term *industrial engineering,* as far as can be determined, was first used publicly when James Gunn suggested in *The Engineering Magazine* that a profession and curriculum of industrial engineering be organized similar to that of electrical and mechanical engineering (Gunn 1901). Going prepared a bibliography of industrial engineering publications and articles which also appeared in *The Engineering Magazine* in 1904. Going had recently written a book titled *Principles of Industrial Engineering*. Other texts on industrial organization and management also appeared about this time: *Factory Organization and Administration* by Diemer in 1910, and *Principles of Industrial Organization* by Kimball in 1913. Both of these men also attended the ASME meeting in 1912, as did Going.

All of the committee members who presented the report and nearly all of those who discussed it were engineers. The profession of industrial engineering as well as the scientific management movement as a whole grew out of the ideas of engineers. By 1912 (and for many years thereafter), these concepts were developed and applied by engineers and presented in engineering publications.

References

Babbage C. 1832. *On the Economy of Machinery and Manufactures*. Carry & Lea. Philadelphia. (Reprint of English Version.)

Barnes, R. M. 1968. *Motion and Time Study*. Sixth Edition. John Wiley & Sons, Inc. New York.

Consiglio, M. 1969. Leonardo da Vinci: the First IE? *Industrial Engineering*. January. American Institute of Industrial Engineers. Norcross, Georgia. 1:71.

Going, C. 1904. Index to the periodical literature of industrial engineering magazine. *The Engineering Magazine*. July. The Engineering Magazine Company. New York. 27(4):643-658.

Gunn, J. N. 1901. Cost keeping: a subject of fundamental importance. *The Engineering Magazine*. January. 20(4):708.

Mechanical Engineering. 1944. July. American Society of Mechanical Engineers. New York. 66(7):491.

The present state of the art of industrial management. 1912. *Transactions of the American Society of Mechanical Engineers*. American Society of Mechanical Engineers. New York. 34:1131-1229.

Schultz, A., Jr. 1970. The quiet revolution: from scientific management to operations research. *Engineering: Cornell Quarterly*. Winter. Cornell University, College of Engineering. Ithaca, New York. 4(4):2-10.

Transactions of the American Society of Mechanical Engineers. 1907. American Society of Mechanical Engineers. New York. 28:33.

Urwick, L. 1956. *The Golden Book of Management*. Newman Neame Limited. London,. p. 10.

2. Background for the Profession

Appropriately, the story of the birth and development of industrial engineering begins with a brief review of the Industrial Revolution. In the period from 1769 to 1800, textile machinery inventions caused the transfer of manufacturing from the cottage to the factory. These inventions, to cite standard texts, were:

1. the spinning jenny consisting of multiple spindles, invented by James Hargreaves in 1770;

2. water power applied to a composite machine including the spinning jenny and other machinery, invented by Richard Arkwright in 1771;

3. the spinning mule, a hybrid machine superseding the spinning jenny and water frame, invented by Samuel Crompton in 1779; and

4. the power weaving loom, invented by Edmund Cartwright in 1785.

At the start of the Industrial Revolution the steam engine was patented by James Watt and introduced in a Nottingham, England, cotton spinning mill in 1785. The manufacture of the new textile machines and the steam engine required new machine tools and manufacturing methods. Thus a new factory system began. One of the first plants was the Soho Engineering Foundry near Birmingham, England, formed in 1796 by Matthew Boulton and James Watt to manufacture steam engines. The Industrial Revolution spread rapidly throughout England. Towns such as Manchester, Glasgow, Liverpool, Nottingham, and Birmingham developed into great industrial areas. The tremendous development of industry made England the dominant nation of Europe for more than half a century.

The first American cotton mill, Slater's Mill, opened in 1793 at Pawtucket, Rhode Island. It was built by Samuel Slater, who had trained under Arkwright. Although a Frenchman named Honore Blanc developed the concept of interchangeable manufacture, it remained for two Connecticut Yankees, Eli Whitney (1798) and Simeon North (1799) to put this system of manufacture into effect. Nevertheless, American industry did not begin to grow and expand until after the Civil War; by 1890 the rate of industrial development had become exceedingly rapid. Shops and mills became large manufacturing organizations, many with subsidiary plants around the world.

The Scientific Management Movement

Frederick Taylor began experimenting in metal-cutting and making time studies in the machine shop of the Midvale Steel Company in 1881. He later applied the scientific approach to simple manufacturing operations such as moving pig iron and shoveling. His paper, "Shop Management," published in 1903 in the *Transactions of the American Society of Mechanical Engineers* set forth the principles that later became known as "scientific management." Most of the work of the early industrial engineering pioneers until 1912 was at the shop level because interest was in the design of machines and processes and the tasks involved in operating them. However, even in these early days, attention spread beyond the shop level.

In France, Henri Fayol extended management theory to the whole enterprise and its organization. Taylor developed theories of "functional foremanship" that divided the organization into planners and operators. Frank Gilbreth wrote the handbook, *Field System,* which included rules, procedures, suggestions, and a system of accounting without books, for supervisors in the building trade. Lillian Gilbreth, throughout her lifetime, emphasized the importance of awareness of the human relationships involved in the construction industry as well as in manufacturing. Henry Towne, as early as 1886, presented the paper "The Engineer as an Economist," (*Transactions of the American Society of Mechanical Engineers* 1886) in which he indicated that the management of works was as important as engineering and deserved a distinct literature and organization of experience. Walter Rautenstrauch wrote a book some years later titled *The Design of Manufacturing Enterprises* (Rautenstrauch 1941). All of these engineers contributed to broadening scientific management concepts beyond the shop level.

The term *scientific management* was given authenticity when used by Louis Brandeis in hearings before the Interstate Commerce Commission. Brandeis, an attorney for shippers who were opposing a proposed freight rate increase, called a meeting of engineers in Henry Gantt's apartment where they agreed that scientific management was a suitable term for the new philosophy of management that they were developing.

In summary, engineers contributed much to the scientific management movement that gave rise to the industrial engineering profession.

The Shoveling Experiment

Two early experiments point to the significance of the unique approach of scientific management: the shoveling experiment and the bricklaying experiment.

Shortly after Taylor went to work for the Bethlehem Steel Corporation in 1898, he began studying the science of one of the oldest jobs in the world: shoveling. His intention was "...the substitution of exact scientific investigation and knowledge for the old individual judgement or opinion," (Taylor 1947) that is, to apply science to an operation in industry. Man is not naturally a scientific, objective individual and so Taylor stated both management and labor would need a mental revolution if industry was to adopt scientific analysis. Applied science can be revolutionary, as was the case with the shoveling experiment.

The key step was to determine the appropriate weight of material that an average man should be able to handle, hour after hour, without undue fatigue--a measure never before determined. Taylor found that 21.4 pounds of material on a shovel would result in a maximum amount handled per day. Apparently, this figure has not been verified, because there is no record that Taylor's experiment was ever repeated. However, the results of this one scientific experiment meant that workers would need a different size shovel for each task depending upon the density of the material they were shoveling, whether it be coal, scrap iron, or wood shavings. Taylor also studied the shoveling methods to find those that would use the fewest motions and involve the least amount of fatigue. In effect, he set a standard for the amount of material to be lifted per shovel for the method used.

This led to recognizing the importance of properly selecting the men to carry out these methods and training them in the recommended methods. Taylor found that workers in the Bethlehem Steel Works were wasting a great deal of time moving from one job to another because the storage yard was one-half to two miles long and almost one-half mile wide. To control the wasted time, Taylor made a map of the steel yard for planning jobs in advance. Instruction cards were given to the workers with notes indicating the types of tools that would be needed and a central tool house was set up to maintain and check the tools.

These improvements made because of this one experiment had the following revolutionary effects on the organization of manufacturing:

— Manufacturing developed the need for industrial engineering departments to develop techniques of time and motion study to set standards for output, methods, tools, and machines.

— Tool and storage rooms became service departments, and later inventory and ordering systems were developed.

— Personnel departments became necessary for writing job descriptions and systematically selecting the workers required in each job.

— Training departments were needed to instruct new workers in the standard methods.

— The layout of manufacturing facilities became important for ensuring that the movement of materials and men were minimized and the new departments were properly located.

— Departments for organizing and planning work and for checking up on day-to-day production (today's production control department) were needed.

— Payroll systems needed reworking to allow for incentive payments to workers who exceeded standard outputs.

Thus the result of applying science to industry was a revolution in the organization of the manufacturing establishment. It also revolutionized instruction in engineering colleges and universities which now had to provide manpower for new fields of work such as industrial engineering and management, cost accounting, personnel management, plant design, and material handling. Courses in each of these fields were developed as they became new areas for professional concentration. Consultants trained by Taylor showed industry how to install these new systems of management.

The final test of any system is "Does it pay?" In this particular experiment at Bethlehem Steel, the crew of four hundred to six hundred men who had worked in the steel yard was reduced to one hundred and forty men who were able to handle the same amount of work, several million tons of material per year. The old system, or rather lack of a system, had cost seven to eight cents per ton. After paying for all clerical work, the time study instruction, and the building and operation of the labor office and tool room, the cost of handling a ton of material was brought down to three to four cents per ton, an estimated annual saving at this particular plant of some $78,000 a year. Unnecessary work was eliminated and men were released for more productive activity. Those continuing on their jobs got more money for less effort and eventually shorter hours. They worked "smarter, not harder" (Drucker 1968).

The Bricklaying Experiment

Gilbreth's application of the scientific method to the laying of bricks had results that were as revolutionary as those of Taylor's shoveling experiment. In 1885, when Gilbreth began working as a bricklayer, he found that the workers did not always use the same motions to lay bricks. Some even used three different methods: one when working on a routine basis, another when working fast, and another when showing the method to someone else. Gilbreth, like Taylor, felt that there should be a standard method, determined systematically, which was prescribed by management, not by the opinions of individuals.

Gilbreth's approach was to analyze each element involved in bricklaying and to eliminate unnecessary motions. Gilbreth reduced the motions from eighteen to four-and-one-half per brick, the half of a motion being accomplished by scraping the mortar from two bricks at a time. A scaffold was invented that would move up as the wall was built higher so that the worker would be laying the brick at the proper level each time and not have to stoop and lift his two-hundred-pound body every time he reached down to pick up brick or mortar. The movable scaffold which keeps work at the proper level is used in many jobs today.

A bricklayer often selects different shades of brick to give a wall a non-uniform appearance. This mixing of shades can be done by a lower-paid worker, but a more recent innovation has been to transfer this task back to the factory where the bricks are randomly palletized. Then this task will already be done for the bricklayer and the finished wall will have a non-uniform appearance. Today, packages of bricks in wire straps are carried up the scaffold and set beside the bricklayer. These post-World War II innovations comply with Frank Gilbreth's principles.

Just as Taylor felt that there should be a handbook to help in the standardizing of data for motion elements in jobs common to many industries, so Gilbreth thought there should be a department of standards in the United States Government that would assemble data for standardizing the trades. He felt that such data need not be confidential but should be passed on like United States Bureau of Standards material and product specifications in other fields. Gilbreth's work, beginning with the study of bricklaying, began to accomplish this, although not under government auspices.

Gilbreth also proposed using unskilled workers for less important jobs, leaving the skilled bricklayer to perform the more expert elements of the work. This concept, involving a modification of Adam Smith's concepts of the division of labor, had been discussed by Charles Babbage in his book *On The Economy of Machinery and Manufactures* in which children or women performed the less skilled jobs at a pin manufacturer and received less pay than men who performed more skilled work. (This concept was "re-invented" recently with a proposal that draftsmen and technicians could do many of the steps in design engineer's work, and do them for less money while the engineer concentrated on the original aspects of designing.)

The work system principles suggested by Taylor and expanded by Gilbreth involved a study of a whole operation, not each workplace as a single entity, although each workplace was analyzed in relation to what went on before and after. Consideration was given to the flow of work into and out of the workplace. For example, in the bricklaying study, the assembling of the bricks and mortar, the mixing of the mortar, and the delivery to the bricklayers, with all the material handling devices involved, were part of the job design.

Gilbreth began bricklaying in 1885 and published the book *Motion Study* in 1911 which dealt mainly with bricklaying. In this book, he described three variables involved in any job:

1. Variables of the worker
2. Variables of the surroundings
3. Variables of motion

Consideration of the variables of the surroundings and of the worker has become the subject matter of a new study for the physiologist, psychologist, and industrial engineer called "human factors." Many academic courses studying operator response to data and stimuli and the motions involved in such response have been introduced since World War II. Much more data of a physiological and psychological nature is available to today's industrial engineer than was available to Gilbreth at the time of his studies.

Gilbreth emphasized the happiness of the worker as well as his productivity both on and off the job. Frank Gilbreth, Jr. suggests that the objective of industrial engineering should be to provide *Time Out for Happiness*, the title of his book about his parents' ideas and contributions (1970).

References

Babbage, C. 1832. *On the Economy of Machinery and Manufactures*. Carry & Lea. Philadelphia. (Reprint of English version.)

Drucker, P. F. 1968. *The Age of Discontinuity*. Harper and Row. New York. p. 272.

Gilbreth, F. 1911. *Motion Study*. D. Van Nostrand Company. New York.

Gilbreth, F. B., Jr. 1970. *Time Out for Happiness*. Thomas Y. Crowell Co. New York.

Rautenstrauch, W. R. 1941. *The Design of Manufacturing Enterprises*.. Pitman Publishing Corp. New York.
Taylor, F. W. 1947. Testimony before a special house committee, 1912. Reprinted in *Scientific Management*. Harper and Brothers. New York. pp. 50-66.
Transactions of the American Society of Mechanical Engineers. 1886. American Society of Mechanical Engineers. New York. 7:428-439.
Transactions of the American Society of Mechanical Engineers . 1903. American Society of Mechanical Engineers. New York. 24:1337-1480.

3. Industrial Engineering Pioneers

The contributions of the pioneers of the industrial engineering profession made many big steps toward modern day techniques. Taylor and the Gilbreths took these big steps toward what was to become industrial engineering. A look into each of their lives and other leaders in the profession will help in understanding how and why each of their contributions came to be.

Frederick W. Taylor

Frederick W. Taylor (1856-1915) graduated from Phillips Exeter Academy in New Hampshire. Eye trouble kept him from attending Harvard as he had planned. Instead, he went to work for the Midvale Steel Company in 1878. While there, he earned a mechanical engineering degree through evening study from the Stevens Institute of Technology. He was a hard worker and also enjoyed sports. His tennis team once won the national doubles championship. Taylor also invented a golf putter which was too good to be permitted for use under golfing rules (Nelson 1980).

Taylor's ideas on employee labor have been criticized, but for his day, they were ahead of the time. He believed that each worker should be given a chance to progress as high and as fast as he could in his job. He believed that workers should have the chance through piecework and wage incentives to earn twenty-five percent or more over standard pay by increasing their output, and do so without fatigue. Taylor did not propose a speedup of work as he was accused of by one labor union; on the contrary, he sought to find a means to determine how much a worker should produce and he thought it scandalous that management did not have this knowledge.

In 1881, at the Midvale Steel Company, Taylor used a stopwatch to determine the proper output per man after analyzing and defining their jobs. In this same year, he obtained permission to study the application of science and scientific research to metal cutting. This latter experiment, continued at other companies, lasted some twenty years at a cost of about $200,000 and drew criticism for not being an example of scientific research. In this same experiment he determined the twelve factors that affect the amount of metal that could be removed per unit of time; then holding eleven of these factors constant, he studied the twelfth. This is why the experiment took so long; with modern statistical methods for the design of experiments, this work could have been greatly shortened, but, for his day, his research methods projected a new concept. This study led to the development of mathematical relationships between the factors.

From the metal cutting experiment, the search for a means to set work standards, and his observations of management's practices, came Taylor's formulation of what he called the principles of scientific management. In reference to the new duties and responsibilities of managers who are guided by these principles, he wrote:

> *First*. They develop a science for each element of a man's work, which replaces the old rule-of-thumb method.
> *Second*. They scientifically select and then train, teach, and develop the workman, whereas previously he chose his own work methods and trained himself as best as he could.
> *Third*. They heartily cooperate with the men so as to ensure that all of the work being done is in accordance with the principles of the science which has been developed.
> *Fourth*. There is almost an equal division of the work and the responsibility between the management and the workmen. The management takes over all work for which they are better fitted than the workman, while in the past, almost all of the work and the greater part of the responsibility were thrown upon the men (Taylor 1911).

Taylor best described the significance of these principles, by saying that they involved a "mental revolution" on the part of both management and labor. To quote his statement:

> Now, in its essence, scientific management involves a complete mental revolution on the part of the working man engaged in any particular establishment or industry—a complete mental revolution on the part of these men as to their duties toward their work, toward their fellowmen, and toward their employers. And it involves the equally complete revolution on the part of those on the management's side—the foreman, the superintendent, the owner of the business, the board of directors—a complete mental revolution on their part as to their duties toward their fellow workers in the management, toward their

> workmen, and toward their daily problems. And without this complete mental revolution on both sides scientific management does not exist (Taylor 1947).

Scientific management was eloquently defined by stating what it is not in Taylor's testimony in 1912 before the Special Investigating Committee of the House of Representatives:

> Scientific management is not any efficiency device, not a device of any kind for securing efficiency; nor is it any bunch or group of efficiency devices. It is not a new system of figuring costs; it is not a new scheme of paying men; it is not a bonus system; it is not a premium system; it is no scheme for paying men; it is not holding a stopwatch on a man and writing things down about him; it is not time study; it is not motion study nor an analysis of the movements of men; it is not the printing and ruling and unloading of a ton or two of blanks on a set of men and saying, "here's your system; go use it." It is not divided foremanship or functional foremanship; it is not any of the devices which the average man calls to mind when scientific management is spoken of. The average man thinks of one or more of these things when he hears the words "scientific management" mentioned, but scientific management is not any of these devices. I am not sneering at cost-keeping systems, at time study, at functional foremanship, nor at any of the new and improved schemes of paying men, nor at any efficiency devices, if they are really devices that make for efficiency. I believe in them; but what I am emphasizing is that these devices in whole or in part are not scientific management; they are useful adjuncts to scientific management, so are they also useful adjuncts of other systems of management (Taylor 1947).

Some of those who helped Taylor install his principles in industry have said that something more than a list of principles or a discussion of a mental revolution was needed to understand his ideas. They could best be explained, they said, by understanding how his ideas were applied in industry. In training his assistants, Taylor emphasized the fallacy of considering parts without the whole; adequate results could not be expected from a plan consisting of one or two phases of scientific management.

As a practical matter, however, Taylor had told his assistants that in many companies not willing to adopt the complete system of his scientific management practices that it might be necessary to select one or two independent operations to show the financial savings that could be made. But the optimum way, the logical manner to introduce scientific management to a company was explained in 1920 by H. K. Hathaway, who was one of Taylor's closest associates. Thirteen steps were involved in the introduction of the whole system:

> 1. General plan of organization, designing departments and subdivisions thereof, defining authority, the nature, scope, and limitations of their activities; and the "relations and the responsibilities of each department to others.
> 2. New plan for physical rearrangement of departments and equipment in accordance with this general plan of organization. Equipment of services used by numerous departments should be centralized in one shop, for example, to avoid duplication of equipment not fully utilized by one department.
> 3. The collection of data relating to the products.
> 4. The collection of data relating to the machinery and other plant equipment.
> 5. Standardization of machines and maintenance.
> 6. Standardization of tools and toolroom.
> 7. The development of the storage system determining minimum and ordering quantities for each article.
> 8. Development of the order system.
> 9. Modification of the time-keeping system to keep records of output, up-to-date records of idle machine time, records of employee incentive earnings.
> 10. Determining the routing system including the complete planning in advance of the work to be done on a product.
> 11. The establishment of and use of the mechanism for follow-up and control of work in progress.
> 12. Time study, the improvement and standardization of methods, tools, machines, and materials. This also involves setting up standard working conditions.
> 13. Cost accounting system. . . [Taylor was a pioneer in the development of modern industrial accounting (Hathaway 1920.]

Work measurement, time study, and methods do not appear until step twelve. Work measurement requires a foundation of sound management practices; it was not the entire scope of Taylor's concepts. His system of management covered the whole manufacturing enterprise, including today's industrial engineering activities.

In his testimony of 1912, and in his writing, Taylor did not claim to be the sole originator of the scientific management movement. In fact, he claimed that his ideas were not new but had been gathered from various sources. This may be true, although no one before Taylor had conceived the procedures for determining how work should be done and the expected outputs; it was his development of systematic methods for doing this that eventually led to a new profession.

Taylor, however, had certain blind spots. For example, his relationship with labor was not a good one because labor misinterpreted his ideas as being that of a speedup of work. Industrialists adopting time study often would change the output standards so that the worker would receive no benefit from an increase in output. This was in direct opposition to what Taylor was trying to achieve: an incentive bonus of at least twenty-five percent for the worker.

Taylor was the first to suggest that rest periods helped improve productivity and he made allowances for them in his time studies. He felt that labor unions were not necessary and that the problem of "soldiering" on the job could be eliminated by the mental revolution that he hoped scientific management would bring about. Taylor found that management was often more difficult to deal with than labor. Management did not want to understand or adopt his new ideas.

Frank Gilbreth

Frank Gilbreth (1868-1924) had planned to attend the Massachusetts Institute of Technology, but had to go to work due to his father's death. He started as a bricklayer's apprentice at seventeen, went into business for himself at twenty-six, and became one of the best known building contractors in the world, with branch offices in New York and London. One reason for his success in winning building contracts was an ability to build quickly; for him, time was of the essence and it was inevitable that he would be interested in time study.

In 1885, on his first day on the job, Gilbreth noticed that the bricklayer assigned to instruct him did his work, as earlier mentioned, in three different ways. The bricklayer was insulted when Gilbreth tried to tell him this. Looking around the construction site, Gilbreth noticed that each of the other bricklayers had his own method of laying brick also. At that time Gilbreth conceived the idea that there should be one best way to do the work and he made it his goal to find that way.

Unlike Taylor, Gilbreth got along with labor; in fact he believed in labor unions. His outgoing personality helped him explain scientific management concepts in training sessions. He became one of the best university and public speakers in favor of the Taylor methods of scientific management.

The title for the book *Time Out for Happiness* (Gilbreth 1970), the third book about Frank and Lillian Gilbreth written by their children, originated from Frank's invention of "happiness units." By performing a job in the best way it could be done, workers could gain happiness units to be spent in leisure rather than in doing unnecessary work. Gilbreth thought workers should be happy in their jobs as well as in their leisure time and that this would be possible through scientific management.

Gilbreth designed flow process charts in which all work in a plant is divided into four types: operations, transportations, storages, and inspections. These charts, still used today, help industrial engineers visualize the entire manufacturing process from raw material to finished product before analyzing each step. Gilbreth outlined the procedures for analyzing each step differently from Taylor. Gilbreth's procedures made it possible to apply science more precisely in the analysis and design of work.

Gilbreth divided the motions of the industrial worker into eighteen and coined the word *therbligs* for these elemental steps of work. (The word "therbligs" is Gilbreth spelled approximately backwards. It was called this to keep others from claiming the concept.) Lillian Gilbreth later stated that she and Frank wished they had assigned time values to each therblig. The therblig "reach," for example, as part of a predetermined time system, has since been assigned a definite amount time per movement of the hand, per given distance.

Some problems have no remedy. Neither Gilbreth nor Taylor could solve the practice of "soldiering" on the job. Today, bricklayers do not lay more bricks per hour than they did when Gilbreth started his work. They follow their own pace, not the pace that Taylor and Gilbreth felt management should determine as appropriate. Gilbreth and Taylor were also not able to alleviate the problem of boredom caused by routine work. However, by applying their concepts to industrial management, it became possible to achieve the same production of goods in lesser amounts of time. Much of this time savings has been passed on to workers in terms of shorter hours. When Taylor started work in the steel mills workers were required to put in ten-hour-plus days. Advocates of the task system of scientific management suggested that shorter working hours would make it possible for child factory workers to play in daylight which they would not otherwise see. In those days, children went to work in factories before daylight and came home after dark. If routine work could not be avoided, it could be minimized. Although industrial engineers today try to design jobs for the benefit of the worker by combining elements and types of work to make jobs more interesting, a residue of routine work remains.

Frank Gilbreth met Frederick Taylor in 1907. He later acted as a consultant to the New England Butt Company when he installed the "Taylor system." A comparison of Gilbreth and Taylor shows that neither attended college as a full-time student. (It is probable that their originality might have been stifled by the routine, academic learning environment of universities.) Both were members of the ASME. (Taylor later served as its president.) Gilbreth was twelve years younger than Taylor and recognized that Taylor's highly significant ideas had come first.

Lillian Gilbreth

Lillian Gilbreth's lifetime (1878-1972) spanned the entire period of the birth, growth, and maturation of the industrial engineering profession. She was one of the pioneers of the scientific management movement and, after the other pioneers had died, devoted the rest of her life to promoting its concepts. During her lifetime engineers not only started a new field of engineering, but contributed to the broader field of management theory and practice. In all of this, Lillian Gilbreth became known as the "first lady of engineering" and the "first ambassador of management" (Reynolds 1972). Examples of this joint contribution to engineering and management were her representations at the First Congress of Scientific Management (CIOS) and for the American Institute of Industrial Engineers (AIIE). Upon her husband's death in 1924, Mrs. Gilbreth chose to represent him and preside over CIOS in Prague. Later, as the AIIE representative to the Council for International Progress in Management, the United States' member organization of CIOS, she attended meetings in Athens, Paris, Mexico City, and Rotterdam. In 1912 and 1913, Mrs. Gilbreth's doctoral dissertation, "The Psychology of Management," was published serially in *Industrial Engineering* magazine.

Mrs. Gilbreth received a doctoral degree in psychology from Brown University. She brought to the engineering profession a concern for human relations and her talks emphasized this concern. Although early in her career she was not allowed to attend engineering meetings because she was a woman, she later became an honorary member of ASME and AIIE, and a fellow of AIIE. She refused to accept the Women's Badge in Tau Beta Pi because engineering students at that time excluded women members. Her position no doubt helped to remedy this situation. She was the first woman to receive the Hoover Medal (an award sponsored jointly by the ASME, the Institute of Electrical and Electronics Engineers, and the Institute of Mining, Metallurgical, and Petroleum Engineers) and the first to be elected to the National Academy of Engineering. She was also a member of Phi Beta Kappa and the recipient of several honorary LL.D and Ph.D. degrees.

Frank and Lillian Gilbreth worked as a team. Although Frank Gilbreth's first book, *Motion Study* (1911) appeared with only his name, all following publications acknowledged her participation. Their son, Frank Gilbreth, Jr., co-author of the book *Cheaper by the Dozen* and author of *Time Out for Happiness*, writes that his father and mother were close partners in all scientific management work after their marriage. (The AIIE recognized this in 1961 by establishing the Frank and Lillian Gilbreth Industrial Engineering Award, its most distinctive honor. The Gilbreths were the first recipients.)

Taylor Versus Gilbreth

The relationship between Taylor and the Gilbreths, although cordial at first, deteriorated over time. Mrs. Gilbreth recalled how, on one occasion when Taylor was a guest in their home she asked him, Henry Gantt, and Frank for advice on the improvement of a certain job. However, later when the Gilbreths were guests in Taylor's home, he had also invited some of his constituents who he allowed to criticize Gilbreth's innovations, such as the introduction of micromotion study and the use of motion picture cameras in time studies. Since Taylor himself had not thought of these, his followers felt that they were not appropriate to use in analyzing industrial operations. Taylor, at the instigation of his followers, took some steps to keep Gilbreth from getting certain consulting jobs in this country, but he refused to do the same in Germany and France, contrary to the wishes of his followers. Although he had become skeptical of Gilbreth's innovations, he depended a great deal upon them for his ability to explain his own system and ideas.

From the beginning Gilbreth was impressed by Taylor's contributions and understood their significance. After Taylor's death, Gilbreth continued to have trouble with Taylor's followers and in turn began to criticize some of the aspects of time study procedures just as they criticized his micromotion studies. They said his approach took too long; he said that time study was inaccurate and involved questionable statistical procedures. He may have been referring to the problem of rating an operator since time studies are not always made on someone who works at what might be considered a normal pace.

Taylor and his followers did not realize that Gilbreth had the essence of a method to measure work more accurately. Through Gilbreth's micromotion study have come the predetermined time systems that have minimized the operator rating problem. This controversy is now history. Lillian Gilbreth tactfully summarized the situation by saying in a conversation with author Howard Emerson that Gilbreth and Taylor complemented each other. Gilbreth extended Taylor's techniques for analyzing the motions of a worker by his use of photography for micromotion study and the concept of therbligs.

References

Gilbreth, F. 1911. *Motion Study*. D. Van Nostrand Company. New York.

Gilbreth, F. B., Jr., and E. G. Carey. 1948. *Cheaper by the Dozen*. Thomas Y. Crowell. New York.

Gilbreth, F. B., Jr. 1970. *Time Out for Happiness*. Thomas Y. Crowell Co. New York.

Hathaway, H. K. 1920. Logical steps in installing the Taylor system of management. *Industrial Management*. August. The Engineering Magazine Company. New York. 60(2): 89-96.

Reynolds, N. Z. 1972. Dr. Lillian Moller Gilbreth 1878-1972. *Industrial Engineering*. February. American Institute of Industrial Engineers. Norcross, Georgia. 4(2): 28-33.

Nelson, D. 1980. *Frederick W. Taylor and the Rise of Scientific Managemen*t. University of Wisconsin Press. Madison, Wisconsin. p. 113.

Taylor, F. W. 1911. *The Principles of Scientific Management*. Harper and Brothers. New York and London. pp. 36-37.

Taylor, F. W. 1947. Testimony before the special house committee, 1912. *Scientific Management*, Part III. Harper and Brothers. New York and London. p. 27.

Vaughn, R. C. 1967. *Introduction to Industrial Engineering*. Iowa State University Press. Ames, Iowa.

4. Seven Criteria for Leadership

The Industrial Revolution set the stage for scientific management, but it was another one hundred years before the scientific approach was applied in industry. In the meantime, practical men starting new industries were forced to innovate. The individuals mentioned in this chapter include some of these men. Others mentioned here were pioneers in the scientific management movement. For further reading on any of the following leaders, consult the *Golden Book of Management* (Urwick, 1956) or *Classics in Industrial Engineering* (Ritchey 1964).

The following seven criteria are used to appraise the importance of early leaders' contributions in the new scientific management movement (see table 2). The criteria were derived from University of Tennessee graduate class seminars at which these early industrial engineering leaders were discussed. The ratings in table 2 represent agreement, although there were differences of opinion.

1. **Originality of ideas**. Did they contribute and develop new concepts?

2. **Followers generated**. Is there evidence that others were persuaded to accept and apply the new concepts?

3. **Publication and promotion**. Did they spread the word?

4. **Academic support**. Were the concepts introduced to academic curricula used to awaken and educate students?

5. **Professional societies**. Were activities related to the new concepts organized and participated in?

6. **Application by industry**. Is there evidence that industry accepted the new concepts?

7. **Judgment of history**. Is the leader's presence still alive in society's memory?

TABLE 2.

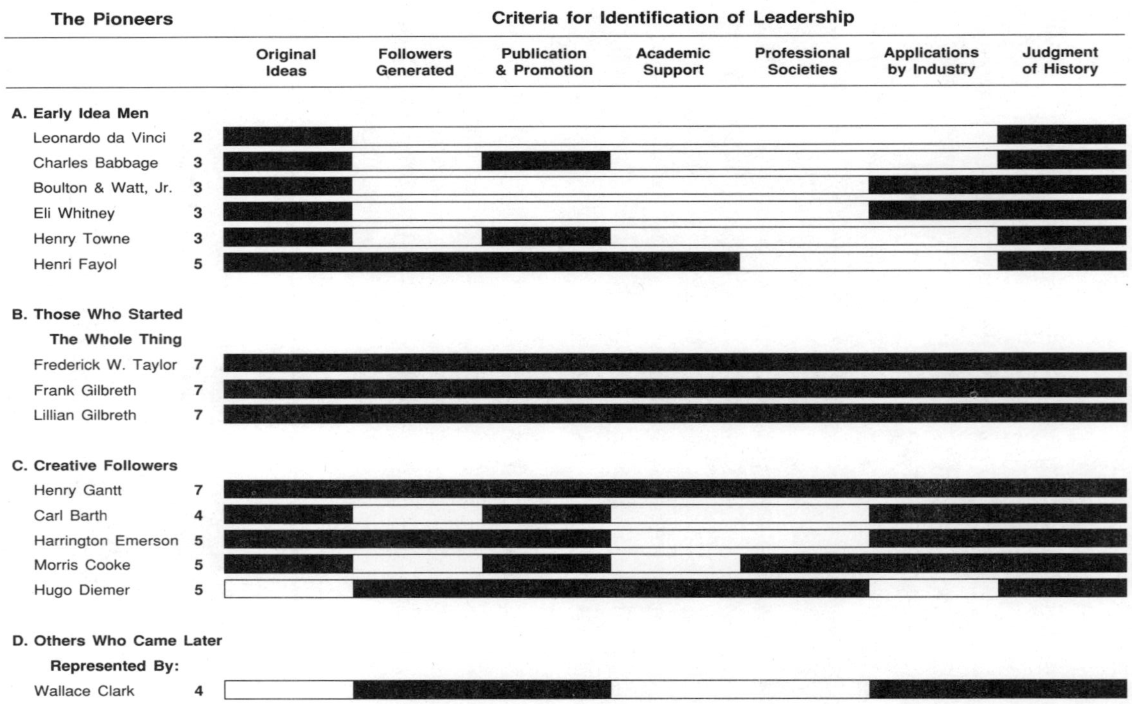

The Pioneers
Criteria for Identification of Leadership
Original Ideas
Followers Generated
Publication & Promotion
Academic Support
Professional Societies
Applications by Industry
Judgment of History
A. Early Idea Men
Leonardo da Vinci 2
Charles Babbage 3
Boulton & Watt, Jr. 3
Eli Whitney 3
Henry Towne 3
Henri Fayol 5
B. Those Who Started The Whole Thing
Frederick W. Taylor 7
Frank Gilbreth 7
Lillian Gilbreth 7
C. Creative Followers
Henry Gantt 7
Carl Barth 4
Harrington Emerson 5
Morris Cooke 5
Hugo Diemer 5
D. Others Who Came Later Represented By:
Wallace Clark 4

Charles Babbage

Charles Babbage systematically recorded ideas on managing industry by scientific methods, but did not apply them. Still, he must be respected for his significant insight in the science of manufacturing. His book, *On the Economy of Machinery and Manufactures*, first published in 1832, included original ideas on division of labor, organization charts, and labor relations. For example, he suggested dividing work into elements and paying less for the less important tasks. He recommended that workers and employers recognize common interests, and that employees recognize that machines would mean more jobs, not less, in the long run. Babbage is also credited with being the first to attempt to build a machine that was a forerunner of the modern computer. He has a recognized high place in the history of mathematics, science, and management, and deserves consideration as one of the early idea men of industrial engineering.

Matthew Robinson Boulton and James Watt, Jr.

In 1796, Matthew Robinson Boulton and James Watt, Jr. became partners in an engineering foundry in Great Britain which manufactured steam engines previously patented by James Watt, Sr. According to one writer (Roll 1930), these two men practiced almost every scientific management technique, including the following:

1. **Market research and forecasting** as the basis for establishment of new business. (Pumping water from mines, operating new textile machines, and transportation opportunities were investigated as prospective enterprises.)
2. **Planned site location** with provisions for adequate communications by land and water, and for possible extension of buildings.
3. **Planned machine layout** in terms of work-flow requirements.
4. **Production planning.**
5. **Machine-operating standards**. (See table 3.)
6. **Standardization of product components**.
7. **Advanced control records** including cost accounting procedures such as maintaining standard books so costs and profits could be calculated on each engine manufactured, and for recording the profits and losses of each department.
8. **Worker training**.

9. **Advanced division of labor**.

10. **Work study**, including development of formulas for standard outputs and as a basis for incentives to exceed them (see table 3).

11. **Payment by results** based on work study using past records, not systematic analysis.

12. **Provision for personel welfare**, with a sickness benefit plan administered by an elected committee of employees.

13. **An executive development program.**

Boulton and Watt, Jr., practiced management techniques that a century later were described more systematically as scientific management. The greatest shortcoming of these men was the fact that they did not write about their accomplishments, except in letters to their partners and acquaintances that were only recently discovered.

Eli Whitney

After inventing the cotton gin, Eli Whitney introduced the concept of interchangeable machine parts: parts made so accurately that any one part could be substituted for another. A similar development occurred about the same time in France; however, Whitney received a government contract to make muskets by this method thus laying the basis for mass production. He also developed factory organization methods.

Whitney was a systematic planner. In his musket factory, he applied his ingenuity and planning ability to laying out the factory and its equipment in a way that would facilitate and increase production. He designed and built a number of machines capable of being operated by semi-skilled or unskilled laborers, thus creating a rudimentary mass production facility. He then trained laborers to operate the machines. Whitney stayed in close contact with every operation in his plant, helping workmen improve their operations and their machines and staying with them until they were skilled artisans, producing at what he thought to be a reasonable rate.

This was not unlike parts of Taylor's program, except that Whitney's production standards were not determined scientifically as were Taylor's. Whitney seems to have assumed many of management's duties as later advocated by Taylor. He introduced many simplified production methods and extensions of machinery in place of labor; however, like Boulton and Watt, Whitney did not write about his management methods.

TABLE 3.

CALCULATIONS FOR ASCERTAINING PIECE-RATES FOR THE FITTING OF STEAM CASES

(Boulton & Watt Collection)

Diameter of Cylinder	Number of Days	Amount per Theorem	Reduced Amount	Panels	Average
		£. s. d.	£. s. d.		£. s. d.
16	14.2	1 9 11	1 3 3	4	
17 3/8	16.2	1 16 6	1 7 1	4	1 7 0
19 1/4	18.1	2 0 10	1 10 3	4	
20 3/4	19.9	2 4 9	1 13 2	6	
21 3/8	20.65	2 6 6	1 14 5	6	
23 3/4	23.5	2 12 6	1 19 2	6	1 17 2 1/2
25 1/4	25.3	2 16 9	2 2 0	6	
27 3/8	27.55	3 5 11	2 6 7	6	
28 3/8	28.78	3 4 9	2 7 11	6	
29	29.8	3 7 1/2	2 9 8	6	2 9 3 3/4
30 3/4	31.9	3 11 9	2 13 1 1/2	6	
31 1/2	32.8	3 13 10	2 14 9	8	
33 1/3	35.0	3 18 9	2 18 4	8	
34	35.8	4 0 6	2 19 8	8	2 19 1 1/2
36	38.25	4 6 1	3 3 9	8	
40	43.0	4 16 9	3 11 3	8	
42	45.4	5 2 2	3 15 8	8	3 18 7
44	47.8	5 7 5	3 19 9	8	
48	52.6	5 18 4	4 7 8	8	
52	57.4	6 9 2	4 15 8	10	4 15 8
60	67.0	7 10 9	5 10 0	12	
63	70.6	7 18 10	5 17 8	12	5 19 10 1/2
66	79.2	8 18 3	6 12 0	12	

Note: The "Theorem" is: d (diameter) x 1.2 - 5 = time in days.
The rate of wages is 2s. 3d. per day.

The above table shows a number of clerical errors which have been left to conform with the original.

Henry Towne

One of the founders of the Yale and Towne Company, Henry Towne, made a major contribution to the engineering profession in 1886 when he recommended that engineers should also be economists (Towne 1886). He recommended the formation of a new division in ASME to be called the economic section. Its objectives would include the development of a distinct literature and the exchange of experiences in the "management of works." This suggestion was debated in ASME over the years and finally in 1921 a management division was established within ASME.

Henri Fayol

A Frenchman, Henri Fayol was a contemporary of Taylor and is often compared with him. Taylor designed management organizations at the shop level; Fayol from the top down. Fayol recommended a military type organization with an executive reporting to one superior; Taylor proposed functional foremanship with service and supervision available to the workman for each part of his job, such as machine setup, operating speed instructions, standard methods, and so forth. Service departments now provide these services, although through one foreman as Fayol had recommended. For example, industrial engineering departments set work standards which are administered by a foreman.

Fayol respected Taylor's contributions to management theory and practice and his work did not duplicate Taylor's. Fayol's book on the principles of management, *General and Industrial Management*, only became available in English in 1949.

Carl Barth

Carl Barth was Taylor's mathematician and acted as a consultant in introducing the "Taylor system" to many companies. As one of Taylor's first associates, he devised practical methods by which the Taylor principles could be implemented. Barth developed the formula of the twelve variables of metal-cutting described in Taylor's book *Shop Management*. To apply this formula, he developed the "Barth complete feed and speed slide rule." The use of this slide rule was explained in a series of twelve articles published in *Industrial Management* magazine between September 1919 and November 1920 titled "Supplement to F. W. Taylor's 'On the Art of Cutting Metals'." Several of Barth's related articles appeared in such publications as, *Transactions of the American Society of Mechanical Engineers*, *Engineering Magazine*, *Bulletin of the Taylor Society*, and others.

Barth eventually went into private practice as a consulting engineer, working both independently and with Taylor. He also lectured on scientific management at Harvard University and the University of Chicago.

Morris Cooke

Morris Cooke met with Taylor at the home of James Dodge, president of the Link-Belt Company, during the time Taylor was president-elect of ASME. Cooke was invited by Taylor to study the operations of ASME and apply principles of scientific management. Thus Cooke was on his way to becoming one of America's first management engineers.

Cooke was also one of four engineers selected by Taylor to teach Taylor's theories in a time when short-cut efficiency engineers went to work in hundreds of plants with disastrous results. The other engineers selected by Taylor were Barth, Gantt, and Hathaway.

In March 1909, the president of the Carnegie Foundation for the Advancement of Teaching made the comment in a letter to Taylor that no one had ever made an economic study of education. Taylor recommended Cooke for this study. Cooke prepared a report to the Carnegie Foundation on the subject of academic and industrial efficiency, published in 1910. Another of Cooke's outstanding contributions was his participation in the arrangements for the Conference on Scientific Management, sponsored by the Amos Tuck School of Administration and Finance at Dartmouth College under its director Harlow Person. There were 275 participants including Barth, Lillian Gilbreth, and other big names in the scientific management movement. Cooke participated in the formation of the Society for the Promotion of the Science of Management which was later renamed the Taylor Society. He later served as its president.

In 1911, the corrupt politicians of Philadelphia lost an election and the new mayor asked Taylor to be his director of public works. Taylor pleaded failing health and suggested Cooke who, at age 29, became Philadelphia's director of public works. Cooke graduated from Lehigh University with a bachelor of science in mechanical engineering in 1895 and received an honorary doctor of science degree from his alma mater in June 1923. He was an active member of ASME. Cooke rated his accomplishments in the following order:

1. Rural electrification promotion.

2. Inexpensive electrification for homes.

3. Progress in labor-management relations. Cooke contributed to

a more enlightened management and more cooperative unions. He persuaded labor leader Samuel Gompers not to oppose scientific management.

4. Conservation of land and water. He was a friend of D. Lilienthal, director of the Tennessee Valley Authority, but did not share the latter's faith in big enterprise.

5. Scientific management in industry. Cooke installed scientific management under Taylor's auspices in leading plants in the country including those in the clothing, textile, shoe manufacture, and printing industries.

Harrington Emerson

Harrington Emerson was a language teacher turned engineer. He had a flair for helping managers, especially those in the railroad business, to reduce costs, now a major goal of all industrial engineers. He accomplished this by emphasizing the concept of efficiency which he described in a series of articles appearing in *The Engineering Magazine* in 1908-09. These articles were published in book form in *Efficiency As a Basis for Operation and Wages* in 1909.

Emerson is best known for his startling statement that railroads could save one million dollars per day using scientific management. This came about when Louis Brandeis, who at the time was an attorney representing shippers who were opposing railroad rate increases in hearings before the Interstate Commerce Commission, remembered having read Emerson's book. Emerson had studied the effects of efficiency methods at the Santa Fe Railroad and written that "The total amount of preventable material and labor wastes and losses in American railroad operations and maintenance approximates $300,000,000 a year . . ."(1909). Brandeis arranged for Emerson to testify before the ICC in 1910 where his startling statement was made. That statement made headlines across the country, the freight rate increase was delayed, and scientific management was no longer an obscure engineering concept.

Henry Gantt

Henry Gantt was an engineer and consultant and friend of both Taylor and the Gilbreths. His influence helped advance the management movement. His invention of the Gantt Chart had been industry's main method of visually planning interrelated jobs until the introduction of the new techniques of PERT (Performance Evaluation Review Technique) and CPM (Critical Path Method). The Gantt Chart was a revolutionary concept that led to improved planning and production controls. It involved visualization of the plant as a whole, one big system made up of interrelated sub-systems. Current authors of sophisticated, mathematical scheduling techniques frequently rely on a Gantt chart to illustrate and to explain their method.

Gantt enlarged upon the scope of industrial engineering by emphasizing the worker. He recommended the following steps for work management:

1. After careful study, define the task;

2. Teach workers how to do it;

3. Provide an incentive in terms of adequate pay or reduced hours to achieve this task; and

4. Provide an additional incentive to surpass it.

This approach was similar to Taylor's, but Gantt applied his concepts wherever human effort was involved. This has led to engineers working in non-manufacturing areas such as hospitals, banks, and service institutions previously untouched by industrial engineering.

Gantt advocated democracy in industry and emphasized industry's place in society, including its responsibility to produce and distribute needed products and avoid wasting resources. He included the use of only part of a worker's skills, idle machinery, and production delays in "wasted resources."

Gantt fulfilled the following points of the seven criteria of leadership:

1. He added original ideas to those he had learned from Taylor.

2. He had loyal followers such as David Porter, who started a motion study laboratory at New York University, and Wallace Clark, who thought the Gantt chart to be the most revolutionary idea in the scientific management movement.

3. He presented his ideas in books, articles and lectures. The respect he received from industry helped gain acceptance of the concepts of scientific management. In fact, the name "scientific management" was chosen while Gantt, Brandeis, and others pre-

pared for the Eastern Railroad Rate Case (more on this later in this chapter). His thoughtful contributions to discussions of management papers were respected and his own papers appeared in the *Transactions of the American Society of Mechanical Engineers.*

4. His Gantt chart was taught as a technique in academic courses, although no widespread impact on curricula resulted as in the case of the ideas of Taylor and the Gilbreths.

5. He was active in ASME and served a term as its president.

6. Gantt applied the ideas of Taylor and the scientific management movement and added contributions of his own. Perhaps this is why he is sometimes called "the first industrial engineer" (Provost 1961).

7. His ideas are still respected. The AIIE, at its 1959 convention on his one hundredth birthday anniversary, recognized Gantt as "the man who first visualized the scope which industrial engineering should occupy."

Frederick Taylor

All of these men contributed significantly to the scientific management movement. Perhaps if they had researched further or had just written down their concepts and ideas they may have had as much influence as Taylor. To see why Taylor's influence is still felt, we will measure him by the seven criteria for leadership.

1. His original ideas included the application of science to industry; the determination, by new methods, of the work content of a job; and the time the job should take.

2. Taylor had loyal followers: Carl Barth and H. K. Hathaway in industry, Hugo Diemer in education, Morris Cooke in public service to name a few.

3. Taylor published his ideas.

4. Taylor's suggestion to the president of Pennsylvania State College that a department of industrial engineering be started was adopted. Followers of Taylor, such as Barth, taught Taylor's ideas

at Harvard University.

5. In 1911, Gilbreth, Cooke, and others helped to start a society that later became the Taylor Society.

6. Unless new ideas are applied, any movement will die in its tracks. Taylor and his assistants installed his applications in industry, in some cases the complete system, in others only parts of the system.

7. A final criterion for appraisal of historical figures is whether their influence continues over the years. In Taylor's case, the influence remains. His ideas have been preserved in the Taylor Society and other international organizations. *The Golden Book of Management* recognizes Taylor as the "father of scientific management." The continued interest in Taylor's ideas is evident by the publication in 1947 of Harper and Row's book, *Scientific Management*, containing Taylor's original paper, "Shop Management," which was originally published in 1903 in the *Transactions of the American Society of Mechanical Engineers*.

Frank and Lillian Gilbreth

Frank Gilbreth, like Taylor, meets the seven criteria for leadership of a new movement, but his contributions cannot be considered without including those of his wife, Lillian. Their work must be considered together when appraising their importance in the scientific management movement.

1. Frank Gilbreth developed innovative concepts, but Lillian helped to generate ideas and put them in presentable form. Her influence is seen in the expansion of applications beyond motion study into the field of work attitudes; applications for the handicapped; and identifying the human relationships in work, often neglected by purely technical engineers.

2. Both were able to influence others to adopt and carry on their ideas. They accomplished this through meetings and classes in New Jersey, England, Germany, and Czechoslovakia. Mrs. Gilbreth continued these on her own after her husband's death.

3. Publication and promotion of the new ideas of industrial engi-

neering and scientific management were a key to continuation of the movement. Gilbreth's colorful and enthusiastic personality made promoting a natural for him. Taylor recognized Gilbreth's ability to explain the Taylor system's new concepts. But the unremitting, intelligent, persistent presentation of scientific management throughout the world and over a period of forty-six years after her husband's death, made Mrs. Gilbreth a key figure in the success of the profession's growth out of scientific management.

4. Academic support is needed to train students in new professions. Gilbreth was a popular speaker on the college circuit and an active member of the American Society of Engineering Education (ASEE). Mrs. Gilbreth became a full professor on Purdue University's industrial engineering faculty, at the Newark College of Engineering, and at the University of Wisconsin.

5. In 1911, Gilbreth helped to found the Society to Promote the Science of Management. The name was changed on Taylor's death to the Taylor Society.

6. The Gilbreths conducted seminars on their methods and were industrial consultants in this country and in Europe.

7. The final criterion for recognition as founders of a profession is continuity and current interest in historical individuals. At its annual meeting in May 1962 at Atlantic City, the AIIE paid a special tribute to Frank and Lillian Gilbreth by designating that conference as the "Frank and Lillian Gilbreth Conference and Convention." Mrs. Gilbreth and most of the twelve children were there. She was also awarded the Hoover Medal at the ASME winter meeting in 1967.

Lillian Gilbreth was a favorite speaker at AIIE student meetings. Unlike many speakers who show up at the last minute and disappear after their talks, she was always interested in the whole program, especially meeting the students who were to carry industrial engineering into the future.

OTHER PIONEERS

Other pioneers in the advancement of scientific management and industrial engineering who contributed as educators, consultants, or industrialists will be introduced in later chapters. Wallace Clark, who promoted the use of the Gantt chart, is selected as a representative of a second generation of pioneers. Having become a consultant to American industry in the early 1900s, he became enthused by the work of Taylor, Gilbreth, and others, and decided that his career should lie in installing these principles in industry.

Starting in 1926, he spent twelve years in Europe introducing the new principles to industry. Clark wrote a book about the Gantt chart (1938) and numerous management booklets for distribution by his consulting firm. His endorsement of the Gantt chart was largely responsible for its acceptance. After his death, his wife continued his consulting practice.

Of the pioneers who followed Taylor and the Gilbreths, none had original basic ideas to the extent of these three. They were the originators; the other pioneers helped to establish and try out the methods. A few, like Gantt, developed techniques such as the Gantt chart, but the essential concepts for industrial engineering were conceived and promoted by Frederick Taylor and Frank and Lillian Gilbreth. These three fulfilled the seven criteria for leadership.

References

Babbage, C. 1832. *On the Economy of Machinery and Manufactures.* Carry & Lea. Philadelphia. (Reprint of English version).

Barth, C. 1919-1920. Supplement to F.W. Taylor's 'On the art of cutting metals'. *Industrial Management.* September 1919-November 1920. The Engineering Magazine Company.

Clark, W. 1938. *The Gantt Chart.* Sir Isaac Pitman and Sons, Ltd. London.

Emerson, H. 1909. *Efficiency As a Basis for Operation and Wages.* The Engineering Magazine Company. New York.

Fayol, H. 1949. *General and Industrial Management.* Pitman Publishing Corp. London. (Translated by Constance Storrs.)

Provost, R. 1961. The contributions of Henry Lawrence Gantt to scientific management. *Journal of Industrial Engineering.* American Institute of Industrial Engineers. New York. SR15.

Ritchey, J. 1964. *Classics in Industrial Engineering.* Prairie Publishing Company. Delphi, Indiana.

Roll, E. 1930. *An Early Experiment in Industrial Organization.* Longmans, Green andCo. London. p. 165.

Towne, H. 1886. The engineer as an economist. *Transactions of the American Society* of *Mechanical Engineers.* American Society of Mechanical Engineers. New York. 7:428-432.

Urwick, L. 1956. *The Golden Book of Management.* Newman Neame Limited. London.

Transactions of the American Society of Mechanical Engineers. 1903. American Society of Mechanical Engineers. New York. 24:1337-1480.

5. Education in Industrial Engineering

> *Factory Economics*—Factory design, equipment, and organization. Selection and arrangement of machinery for given conditions. Shop plans. Study of organization and management of industrial establishments. Senior, 2nd term.
> Daily, at 9
> Professor Diemer.

So read the catalog description of a course offered by the mechanical engineering department of the University of Kansas for the 1901-1902 term. This was the first industrial engineering and shop management course offered by a college or university.

Other universities gradually added industrial engineering courses. At Cornell University, Dexter Kimball introduced an elective course for mechanical engineering seniors in 1904. He intended to name the course economics of production, but this title was changed to works management. Based on his lectures at Cornell, Kimball wrote a book, *Principles of Industrial Organization*, published by McGraw-Hill, Inc. in 1913. For over forty years it remained a popular text.

In the 1907-1908 academic year, Walter Rautenstrauch introduced a works management course at Columbia University. Its course description reveals that several disciplines were included—methods, layout, organization, working environment, and cost analysis:

> *Business Methods*—methods of manufacture and output affecting systems of management of mills and factories and cost of production. Departmental division and distribution of authority and responsibility as affecting the directness of producing. Departments of engineering, construction, sales and stores. Executive department. Methods and forms of reports to directors and various departments. Labor accounting systems. Cost determination and selling price. Purchase of raw material and sales of products. Utilization of scrap and character of construction with regard to lighting, ventilation, sanitation, insurance, and reduction of fire hazard. Principles of mill construction and power.

The First Industrial Engineering Curriculum

Frederick Taylor recommended to James Beaver, president of the board of trustees of Pennsylvania State College (now Pennsylvania State University) that mechanical engineering be taught from the standpoint of manufacturing rather than from the standpoint of power plant tests and higher mathematics. He recommended Hugo Diemer for the job. Diemer, born in 1870 in Cincinnati, Ohio, had received a bachelor's degree in mechanical engineering from Ohio State University in 1896 and had worked as a production engineer before beginning a teaching career in 1900. Diemer had a likeable personality and a genuine interest in people.

At the University of Kansas in 1901, Diemer taught a course in shop management previous to the publication of Taylor's 1895 paper, "A Piece Rate System."

Upon Taylor's recommendations, Diemer became head of mechanical engineering at Pennsylvania State in 1907 and began exploring the possibility of initiating a degree program aimed at training students to be industrial managers as well as engineers. Soon he proposed and initiated a program in shop management at Pennsylvania State under the auspices of mechanical engineering. The following report from Dean of Engineering John Jackson to Pennsylvania State President Edwin Sparks reveals Diemer's ideas of the responsibilities of the industrial engineering profession at that time (1908):

> A new course; the fact that industrial education for engineers had heretofore confined itself, to quite a large extent, to the more technical side of industries at the expense of the more strictly business departments, is just beginning to be realized. Though it is not proposed in the school to modify unduly the present curricula, it's believed that we can afford to devote some hours to the study of economics, shop management, and the handling of labor without weakening the rigors which this school has heretofore insisted upon. It is thought desireable, in addition to those modifications to the body of established courses, to inaugurate a new course, especially designed to fit men for business positions in industrial establishments including transportation, buying and selling, and with this object in view, Professor Diemer at the request of the School, has made a tentative course, which appears to be a satisfactory basis upon which to work.

Thus, in 1908, the first curriculum in industrial engineering was developed by changing and modifying the existing mechanical engineering curriculum for those students selecting this option. The two-year course was limited to juniors and seniors in mechanical engineering, thus making Pennsylvania State the first institution in the world to offer a baccalaureate program in industrial engineering. Enough interest was shown to begin a separate four-year curriculum in 1909 and Diemer was considered the logical choice to head the program. Consequently, he resigned his mechanical engineering position to accept the new responsibility.

At its inception, the objective of Diemer's industrial engineering curriculum was similar to that of today. In the Pennsylvania State course catalog of 1909-1910, a course in industrial engineering was presented for the first time:

> This course is intended especially to prepare for positions that deal with the side of industrial organizations that has to do with business management, works management, superintendence, purchasing and sales. It deals largely with the application of the sciences and humanistic studies to industrial ends. It prepares for the competent handling of such subjects as the determining of costs, depreciation, statistics, proper distribution of expense, economic production, systems of remunerating labor and raising labor efficiency, the handling and records of stock and orders, sales, purchasing, corporation accounting and allied work.
>
> For the purpose indicated, well-trained engineers are required; hence the course includes all the fundamental engineering work in mathematics, drawing, physics, chemistry, machine design, heat engineering, mechanics, hydraulics, and structures that are common to all other engineering courses. But more time should be and is given to such general studies as modern languages, English, economics, logic, and psychology and specialized work in accounting, factory management, shop time study, machine tools and methods and shop practice in general.

The Pennsylvania State College course catalog for the next year, 1910-1911, listed the following industrial engineering faculty:

Hugo Diemer, M.E., Professor of Industrial Engineering
Frank B. Faulkner, Instructor in Machine Tool Work
Harvey R. Reidenbaugh, Instructor in Forging and Foundry
George H. Resides, Instructor in Practical Mechanics
Herbert W. Small, A.B., Instructor in Woodwork
E. M. Broderick, Instructor in Woodwork
E. L. Bowman, Instructor in Woodwork
John J. Light, Instructor in Machine Shop

Included in the facilities were woodworking shops, a foundry, forge shops, and machine shops. The curriculum was published in volume 35 of *The Engineering Magazine.*

Diemer's perception of the principles of management plus his experience as a consulting engineer led him to write the classic textbook, *Factory Organization and Administration,* which was published in 1910. This first textbook in the field of industrial engineering was an immediate success. Topics covered were: organization and control, centralized planning, bills of material, the drafting department, stores and stock department, aids in taking inventory, the purchasing department, time and motion studies, wage systems, inspection, maintenance and tools, cost accounting, personnel work, office management, and executive reports.

Diemer retained an active connection with industry and taught at LaSalle Extension University in Chicago for the next nineteen years. Most of the students at LaSalle worked in industry during the day and studied at night. Diemer was not satisfied with the available textbooks and so wrote *Foremanship Training,* which outlined the desired characteristics and responsibilities of foremen. It was based primarily on Taylor's ideas.

Diemer died in Chicago at the age of 68. Although he did not originate any new ideas, his outstanding contribution was the development of industrial engineering curriculum and merging it with industry.

Subsequent Curricula

William Kent began an industrial engineering curriculum at Syracuse University in 1908. Although this original program was discontinued, Syracuse later revived an active industrial engineering program. Also at this time, the Harvard School of Business Administration adopted the Taylor system for teaching shop management courses. An industrial engineering program was also introduced at the University of Kansas in 1912.

As mentioned earlier, Rautenstrauch introduced a course titled works management at Columbia University in the 1907-1908 academic year. Charles Going's lectures in this course were published in 1911 in *Principles of Industrial Engineering*, the first book using the words "industrial engineering" to describe the profession. In 1919, Rautenstrauch formed the industrial engineering department at Columbia and continued to head it for some thirty years. His course offering was described by Raymond Villers, a close associate, as advocating the "broad concept" of industrial engineering:

> Much had already been accomplished by the Scientific Management Movement. Great progress had been made in the direction of improving the efficiency of production methods, of plant organization and that of various special activities within the industrial enterprise. No attempt had been made, however, to integrate each and every activity related to industrial production. Such an effective integration is precisely the goal of the broad concept of Industrial Engineering. Today, this concept is best expressed in the AIIE definition of Industrial Engineering adopted in 1956 (revised in 1985) by the Seventh National Convention (Villers 1958).

In June of 1912, Rautenstrauch presented a paper, "Teaching the Principles of Scientific Management," to the twentieth annual meeting of the Society for the Promotion of Engineering Education held in Boston, Massachusetts. Among the other speakers at this session chaired by Frank Gilbreth, were S. E. Thompson, Henry Gantt, Harrington Emerson, H. K. Hathaway, Robert Kent, Hugo Diemer, William Kent, and H. J. Porter. In this paper, Rautenstrauch stated that not only is it important for the student to learn how to design machines, but also that there may be a variety of designs that would perform the same service. Rautenstrauch continued:

> Unless a machine can be built at a cost sufficiently low to meet the demand it will never get beyond the experimental stage and there is absolutely no use in designing a machine if its cost of production or operation exceeds what can be properly paid. It is certainly important to teach the students how a bar can be turned round or a plate made flat or how every necessary step to making a machine may be car- ried out by a mechanic; but it is equally important that he should know their relative costs so that he may avoid the expensive

> operations or reduce their use to a minimum. He should also be taught that the maximum economy in producing that machine will result when it is made in quantity, and why, and that quantity production will involve changes in design, changes in shop processes because it permits the use of special instead of standard tools and finally the most important, that quantity production involves many men and that many men require management that each may produce the maximum for his wages, and that such an organization for economic production is sub ject to laws and principles of far greater consequence than all the thermodynamics that was ever formulated; and yet laws of essentially the same nature as those that control the design of his machine (Rautenstrauch 1912).

Rautenstrauch combined his interests in engineering and the human mind. He is best known for his book and articles promoting use of the breakeven chart, the first mathematical model of a complete industrial enterprise. As early as 1910, he wrote about a solution not unlike linear programming in the article "A Comparison of Lathe Head-Stock Characteristics." Among books written by him were, *The Economics of Business Enterprise, The Design of Manufacturing Enterprises,* Who *Gets The Money,* and with Villers, *Budgetary Control.*

Curriculum Development

The first industrial engineering curriculum contained certain courses from the natural and physical sciences plus a variety of courses from other engineering disciplines. In many cases industrial engineering first appeared as an optional course of study in a mechanical engineering program. Shop courses were available and were included to provide the graduate engineer with a practical background. Textbooks were few. Of those available were Kimball's *Principles of Organization,* D. V. Merrick's *Time and Motion Study,* C. Bertrand Thompson's *Scientific Management,* and Diemer's *Factory Organization and Administration.*

Typical was the situation H. G. Thuesen found himself in when he joined the newly formed department of industrial engineering in 1925 at Oklahoma A & M College (now Oklahoma State University). In his senior year as a mechanical engineering student at Iowa State College, he elected to take the industrial engineering option and studied under J. O. Keller, one of Diemer's first graduates at Pennsylvania State College.

During the summer of 1925, Thuesen applied to a number of

schools and was hired by Edward Boyd at Oklahoma A & M College. Upon his arrival, he found that the department of industrial engineering had been renamed from the department of shop practice. Department head, DeWitt Hunt's main interest was shopwork for engineers and industrial arts education teachers. However, three general industrial engineering courses had been added to the departmental offerings and Thuesen was permitted to alter the curriculum with a shift of emphasis in instruction from manual to analytical skills.

As Thuesen was promoted to acting department chairman and then department chairman, he added such courses as time study, tool design, factory planning, manufacturing processes, industrial organization, scientific management, industrial management, and production planning. The department was accredited by the Engineers Council for Professional Development (ECPD) in 1937.

Thuesen wrote the textbook *Engineering Economy* (1950), designed high accuracy timing devices for the time-and-motion field, designed a memo-activity camera, and was a co-developer of the world's first operable parking meter. (A tribute to Thuesen, often called the father of industrial engineering in the Southwest, appeared in the December 1973 issue of *Industrial Engineering*.)

A report of changes made in the industrial engineering curricula at the University of Pittsburgh during the period 1920 to 1960 was prepared by Walter Turkes of the University of Pittsburgh for the special feature issue of the *Journal of Industrial Engineering* (September-October 1962). Included were the charts that are reproduced in table 4 and figure 1 which depict graphically and in tabular form the contents of the industrial engineering curriculum in five-year intervals.

Courses in the natural sciences and engineering sciences remained in about the same proportion during this period, with a slight increase in the later years. A major trend appeared in 1945, when industrial engineering sciences expanded significantly with a corresponding reduction in business administration courses. In 1955, courses in computer methods, linear programming, advanced statistics, and managerial and engineering economics, all requiring an increase in a mathematical background, appeared. Emphasis on the humanities and social sciences also increased in 1955. By 1960, the number of business administration courses had dropped to zero.

TABLE 4.

Analysis of University of Pittsburgh IE Curricula, 1921-1960 by Course Titles, Five-Year Periods

COURSE TITLES	1921	1925	1930	1935	1940	1945	1950	1955	1960
NATURAL SCIENCES									
Mathematics	x	x	x	x	x	x	x	x	x
Physics	x	x	x	x	x	x	x	x	x
General Chemistry	x	x	x	x	x	x	x	x	x
Differential Equations									x
ENGINEERING SCIENCES									
Mathematical Drawing	x	x	x	x	x	x	x	x	x
Surveying	x	x	x	x	x	x			
Electrical Engineering	x	x	x	x	x	x	x	x	x
Heat Power	x	x	x	x	x	x			
Theory and Applied Mechanics	x	x	x	x	x	x	x	x	x
Design	x	x	x	x	x	x	x		
Steel Manufacturing	x	x	x						
Public Utilities	x								
INDUSTRIAL ENGINEERING									
Engineering Reports	x	x							
Thesis	x	x	x	x	x	x	x	x	x
Time Study	x		x	x	x		x	x	
Production Control	x	x	x	x	x	x	x	x	x
Seminar	x	x	x	x	x	x	x	x	x
Manufacturing Methods		x							
Public Utilities		x							
Research in Industry		x	x	x	x	x	x	x	
Industrial Management			x	x	x				
Industrial Engineering Problems			x	x					
Motion Study					x				
Machine Tools and Processes						x	x	x	
Principles of Industrial Engineering						x	x	x	
Industrial Safety Engineering						x	x	x	
Job Evaluation						x			
Advanced Motion Study						x			
Time and Motion Study						x			
Methods Engineering						x			
Production Standards Deviation							x	x	

COURSE TITLES	1921	1925	1930	1935	1940	1945	1950	1955	1960
NATURAL SCIENCES									
Quantitative and Qualitative		x							
Geology		x							
Mathematics of Investment					x				
ENGINEERING SCIENCES									
Hydraulics	x								
Chemical Engineering Practice		x							
Economics of Contracting		x							
Highway Transport		x							
Metallurgy				x	x				x
Thermodynamics							x	x	x
Design of Mechanics									x
INDUSTRIAL ENGINEERING									
Wage Structure Analysis							x	x	
Engineering Analysis and Economy							x	x	
Industrial Engineering Applications							x	x	
Quality Control								x	
Engineering Computational Methods									x
Manufacturing Processes and Equipment									x
Methods Engineering Analysis									x
Digital Computers Principles and Programming									x
Engineering Statistics									x
Engineered Cost Analysis and Control									x
Productivity Measurement									x
Linear Programming Fundamentals									x
Managerial and Engineering Economics									x

COURSE TITLES	1921	1925	1930	1935	1940	1945	1950	1955	1960	COURSE TITLES	1921	1925	1930	1935	1940	1945	1950	1955	1960
										INDUSTRIAL ENGINEERING BUSINESS									
Industrial Management	x	x								Business Law	x	x							
Controls and Specifications	x									Foreign Trade	x	x							
Sales Engineering	x	x								Accounting		x							
Sales Management	x									Industrial Employment		x							
Transportation	x	x																	
										BUSINESS ADMINISTRATION									
Finance	x	x		x	x					Commerce and Marketing			x	x	x				
Accounting	x		x	x	x	x	x	x		Business Law			x	x	x	x			
Banking	x	x								Transportation			x	x	x				
Statistics	x	x	x	x	x	x	x	x		Foreign Trade			x						
Employment Practices	x									Industrial Relations					x				
Personnel Administration	x		x	x						Labor Problems						x			
										HUMANITIES AND SOCIAL SCIENCES									
English Composition	x	x	x	x	x	x	x	x	x	Political Science								x	
Economics	x	x	x	x	x	x	x		x	Literature								x	
Psychology	x	x	x	x						Lower Level Social Sciences									x
Written English		x								Upper Level Social Sciences									x
Technical English			x	x	x	x	x	x	x	Lower Level Humanities									x
History						x	x			Upper Level Humanities									x
										SHOP									
Foundry	x									Forge	x	x							
Machine	x	x	x	x	x					Heat Treatment	x								
										OTHER									
Engineering Talks	x				x					Current Engineering Topics	x								
Plant Visits	x		x	x	x					Military	x	x	x	x	x	x	x	x	x
Physical Education	x	x	x	x	x	x	x	x	x	Electives			x	x	x	x	x	x	

Source: *The Journal of Industrial Engineering*, September-October 1962, page 372.

FIGURE 1.
Analysis of University of Pittsburgh IE Curricula, By Subject Areas, Five Year Periods, 1921-1960

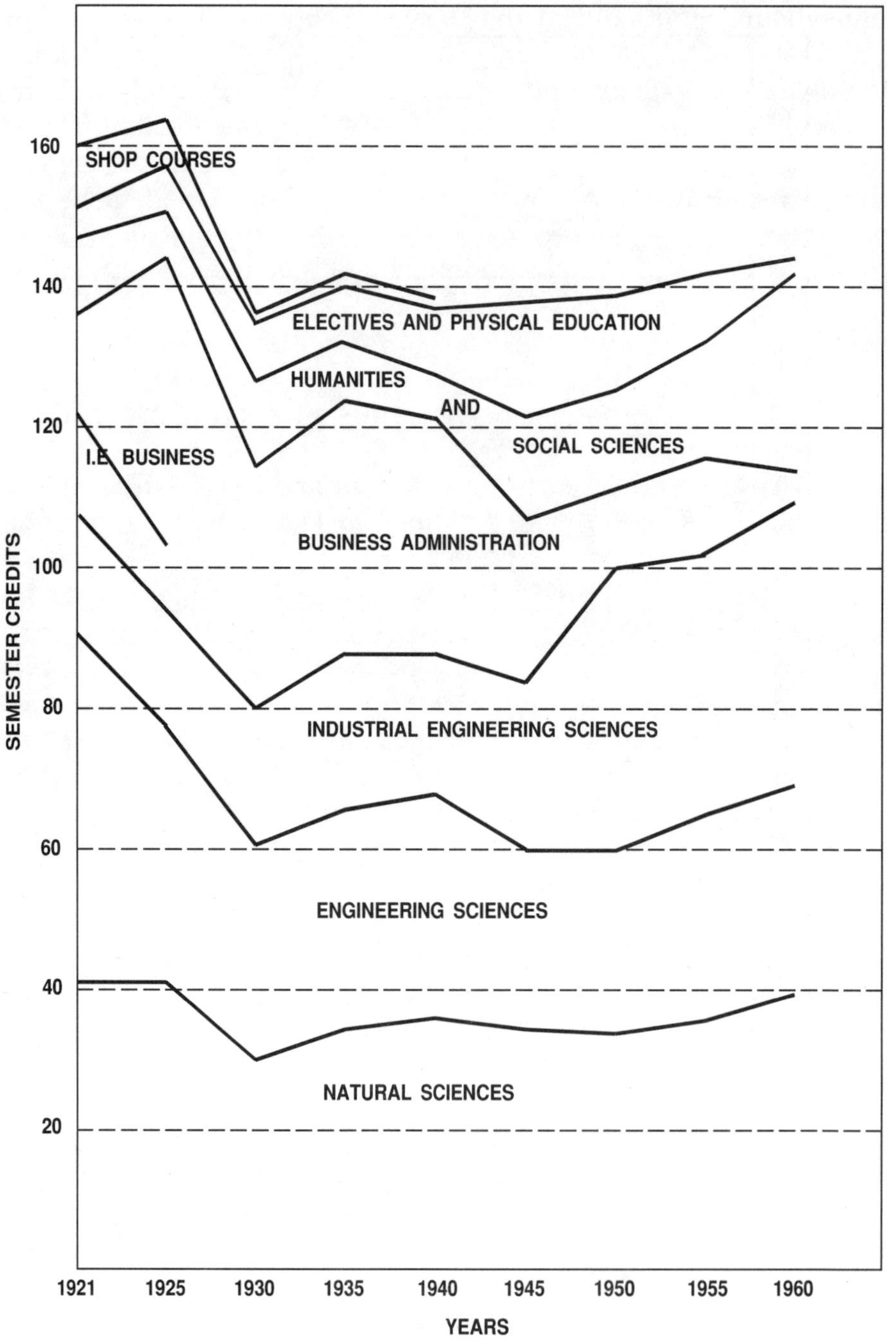

Source: *The Journal of Industrial Engineering.* September-October, 1962. Page 372. © Institute of Industrial Engineers, 25 Technology Park/Atlanta, Norcross, Georgia 30092.

The Industrial Management Connection

While Diemer was setting up the industrial engineering curriculum at Pennsylvania State College, the Harvard School of Business Administration was being organized in Cambridge, Massachusetts. Carl Barth, Hathaway, Thompson, and a number of other Taylor followers tried to convince the new school to teach Taylor's concepts of scientific management. Although Taylor approved of this, he felt that the engineering school was the best place to present the scientific management program and principles. It was valid reasoning, considering that all instructors of the early concepts of scientific management at Pennsylvania State College or Harvard were engineers.

This parallel approach to teaching scientific management concepts in business administration schools and engineering schools continued through the years, at times causing confusion and conflict, at other times with varying degrees of cooperation.

An example of an early approach to handling this dual interest was at the University of Tennessee, where, in 1913, Charles Ferris, dean of engineering, started a course called scientific management. Ferris was a member of ASME and no doubt had read Taylor's papers and attended ASME's national meetings where management programs were conducted. The scientific management course was an elective for seniors in engineering but was required of all business administration students. Three years later, the college of business administration initiated its own courses in the field of scientific management. When the present curriculum in industrial engineering was implemented in 1948, students in the college of business administration's department of industrial management attended courses in motion and time study in the industrial engineering department and special courses in drawing and use of the slide rule in the engineering college. Gradually, the department of industrial management increased emphasis on organization and personnel studies, while the industrial engineering department concentrated on the physical basis of plant operations such as plant layout, motion and time study, and materials handling.

Contributions of a Liberal Arts College

In 1912, the Tuck School of Business Administration at Dartmouth College held a symposium at which nearly all of the pioneers in scientific management were present, including Taylor and the Gilbreths. The meeting report is a source of information about the scientific management movement.

The Thayer School of Engineering at Dartmouth is one of the oldest engineering schools in the country, but limited to civil, electrical and industrial engineering. For many years the Tuck and Thayer schools were on opposite sides of the campus, then later, they located just across the road from each other. The historical relationship between management and engineering education culminated in the Tuck-Thayer Center being built at Dartmouth college to serve ". . . as a bridge between the graduate schools of business administration and engineering at Dartmouth, a recognition of the growing interdependence between the technical and managerial segments of American enterprise" (Trustees Authorize Tuck-Thayer Center 1971). This decision to build the center was stated by Dartmouth President John Kemeny, a mathematician and a pioneer in the teaching of computer technology and operations research.

Although Dartmouth is an undergraduate liberal arts college with limited graduate emphasis, the Tuck-Thayer Center was intended to ". . . facilitate the fertilization of ideas and information among future managers and engineers within the atmosphere of the liberal arts" (Trustees Authorize Tuck-Thayer Center 1971).

Expansion of Industrial Engineering Curricula

Diemer's curriculum in industrial engineering at Pennsylvania State College was formed in 1908. Until 1920, only three more had been added, and in the 1920s, ten more. Between then and 1940, there was a gradual expansion. From 1945 on, expansion of accredited industrial engineering curricula has been spectacular. Table 5 shows the growth of industrial engineering curricula. The table shows the date when an autonomous department was set up with an industrial engineering curriculum. After 1937, when ECPD accreditation started, dates of accreditation are used in the absence of other information.

In 1922, there were nine colleges offering courses in management. They were: Columbia University, Massachusetts Institute of Technology, New York University, Pennsylvania State College, Purdue University, University of Kansas, University of Pittsburgh, University of Tennessee, and Yale University. By 1937, there were seventeen institutions offering a distinct degree as well as a distinct curriculum. Fourteen other institutions had a distinct curriculum but granted a general degree, and thirty-three colleges and universities offered management courses either in engineering or business.

Table 5. Start-up Dates of Industrial Engineering Curricula (United States and Canada)

Date of First Autonomous Department	Institution
1908	Pennsylvania State College
1912	University of Kansas
1919	Columbia University
	New York University
1921	University of Pittsburgh
1922	University of Alabama
	Lafayette College
1924	University of Michigan
	Oklahoma A & M College
1925	GMI Engineering & Management Institute
	Lehigh University
	Montana State University
	Ohio State University
	Syracuse University
1929	Virginia Polytechnic Institute and State University
1931	Cornell University
1932	North Carolina State University (Raleigh)
1934	University of Florida
	Oregon State University
	Texas Tech University
1939	Northeastern University
1940	Illinois Institute of Technology
	Polytechnic Institute of New York (New York University)
1941	Texas A & M University
1942	University of Southern California
1945	Georgia Institute of Technology
	Stanford University
1946	Johns Hopkins University
	State University of New York at Buffalo
1947	University of Miami
	University of Rhode Island
	Rutgers, The State University of New Jersey
	University of Washington

1948	Bradley University
	University of Houston
	University of Massachusetts at Amherst
	University of Tennessee
	Wayne State University
	Wichita State University
1949	University of Arkansas
	Lamar University
1950	Southern Methodist University
	University of Toledo
1952	North Dakota State University
1955	University of Puerto Rico, Mayaguez
	Purdue University
	University of Illinois, Urbana-Champaign
	West Virginia University
1956	Arizona State University
	University of California, Berkeley
	California Polytechnic State University-San Luis Obispo
	Farleigh Dickinson University
	Iowa State University
1958	University of Missouri -Columbia
	Northwestern University
1959	University of New Haven
	Kansas State University
	University of Toronto
	Western Michigan University
1960	University of Arizona
	New Jersey Institute of Technology
	St. Mary's University
	San Jose State University
	University of Texas at Arlington
1962	University of Central Florida
	University of Iowa
	Mississippi State University
	Newark College of Engineering
	University of South Florida
	Tennessee Technological University
1963	University of Oklahoma
	Western New England College
1964	Auburn University
	University of Alabama in Huntsville

1965	Louisiana State University
	Technical University of Nova Scotia
1967	Louisiana Tech University
	Ohio University
1968	Rochester Institute of Technology
	University of Windsor
1969	University of Texas at El Paso
	New Mexico State University
1970	California State University-Fresno
	Cleveland State University
	University of Nebraska
	The University of Utah
	University of Wisconsin-Madison
1972	University of Illinois at Chicago
	Ecole Polytechnique de Montreal
1973	University of Michigan-Dearborn
1974	Southern Illinois University at Edwardsville
1975	University of Cincinnati
1976	University of Wisconsin-Milwaukee
	California State Polytechnic University-Pomona
1977	University of Louisville
	North Carolina Agriculture and Technical State University
1978	Rensselaer Polytechnic Institute
	University of Bridgeport
1983	Clemson University
	St. Mary's University
1984	University of Regina

References

Dartmouth College Conference on Scientific Management. 1912. Dartmouth College. Hanover, New Hampshire. October 12-14.

Diemer, H. 1910. *Factory Organization and Administration*. McGraw-Hill, Inc. New York.

Going, C. 1911. *Principles of Industrial Engineering*. McGraw-Hill, Inc. New York.

Industrial Engineering. 1973. American Institute of Industrial Engineers. Norcross, Georgia. 5(12):7.

Kimball, D. 1913. *Principles of Industrial Organizatin*. McGraw-Hill, Inc. New York.

Rautenstrauch, W. 1910. A comparison of lathe head-stock characteristics. *American*

Society of Mechanical Engineers Journal. American Society of Mechanical Engineers. New York. Vol. 32.

Rautenstrauch, W. 1912. Teaching the principles of scientific management. *Proceedings of the Twentieth Annual Meeting, 1912*. Society for the Promotion of Engineering Education. Part 1, 20:87-88.

Taylor F. 1895. A piece rate system. *Transactions of the American Society of Mechanical Engineers* . American Society of Mechanical Engineers. New York.

Thuesen, H.G. 1950. *Engineering Economy*. Prentice-Hall, Inc. Englewood Cliffs, New Jersey.

Trustees authorize Tuck-Thayer Center. *Dartmouth Alumni Magazine*. 1971. Dartmouth Alumni Association. Dartmouth, New Hampshire. p. 21.

Turkes, W. 1962. *Journal of Industrial Engineering*. American Institute of Industrial Engineers. New York. 13(5):371-377.

Villers, R. 1958. Walter Rautenstrauch's pioneering work - a 50th anniversary. *The Journal of Industrial Engineering*. January-February. American Institute of Industrial Engineers. 9(1):7.

6. Industry's Acceptance and Contributions

While steps were under way in the field of education to promulgate and expand the concepts of scientific management and industrial engineering, many private companies were experimenting with applications of the new ideas. Frederick Taylor applied the principles that he had developed at the Midvale Steel Company during the late 1800s and Bethlehem Steel Corporation retained him as a management engineering consultant from 1899 to 1901. His followers also became consultants to a wide variety of companies. The New England Butt Company hired Frank Gilbreth as a consultant to introduce the Taylor system and cooperated with him in developing his micromotion study. Later, Gilbreth introduced his ideas and principles at such companies as Pierce-Arrow, Remington Typewriter, Eastman Kodak, Cluett-Peabody, Lever Brothers, and the U. S. Rubber Company.

Companies that adopted scientific management principles before 1920, either on their own or with the help of consultants, and that continue to practice industrial engineering functions deserve special recognition in the profession's history. Included among these are the following, whose contributions (based on information supplied by the respective companies) are discussed further in the chapter.

1. Industrial engineering was introduced in the Armstrong Cork Company from 1912 to 1914.

2. The founder of the Dow Chemical Company, H. H. Dow, became interested in motion study in 1917.

3. Eastman Kodak, from the days of Gilbreth's use of photogra phy in motion study, became a leader in the application of tradi tional industrial engineering and a pioneer in human factors studies. Kodak installed an industrial engineering department in 1912.

4. Eli Lilly and Company in Indianapolis started a long history of industrial engineering applications in 1914 by employ ing Harrington Emerson as a consultant.

5. Stockham Valves and Fittings implemented step-by-step applications of industrial engineering.

6. Western Electric Company chose to apply scientific management in an integrated way. Its use of specifications for all materials, supplies, processes, wastes, production, quality, and jobs supported this. It and the American Telephone and Telegraph Company pioneered the use of the Shewhart quality control charts and later, the use of human factor studies. The studies at the Hawthorne plant on incentives provided insight and revised previous concepts.

These companies have been selected to illustrate the contributions of industry to the development of industrial engineering. They were, and continue to be, a part of a cooperative attitude between industry, education, and the engineering fields.

Armstrong Cork Company[1]

Industrial engineering principles were initially adopted at the Armstrong Cork Company from 1912 to 1914 when the consulting firm of Miller, Franklin, and Bassett was hired to investigate Armstrong's manufacturing processes. The result was the installation of a cost system, a quality control plan, and a maintenance planning program.

In 1917, the same consulting firm was retained to install an incentive system in the Lancaster Floor Plant's World War I shell shop. J. A. Cummings, who had supervised the work in the shell shop, was named Armstrong's first full time industrial engineer and in 1927 he assumed the title of company industrial engineer.

Additions to industrial engineering responsibilities came rapidly as Cummings continually demonstrated his success in completing managerial assignments. Under his direction, the company launched its first program of group incentives, an incentive bonus payment plan based on the group effort of a crew of two or more individuals. Armstrong industrial engineers not only pioneered in this type of wage incentive, but also extended it into industries where group incentives had never before been applied.

In the mid-1920s, a tremendous expansion program occurred in the floor coovering operations at Armstrong. During this time Cummings directed Armstrong into several firsts in its list of industrial engineering accomplishments such as:

[1]Material in this section is from information supplied by the Armstrong Cork Company with the permission of William Van Pelt, Manager, Employment and College Relations Department.

1. Corporate hiring of accredited industrial engineers as a means of staffing all plants with professionally trained men.

2. Establishment of production planning as an inventory management function instead of as just a scheduling technique.

3. Origination of package engineering to establish uniformity of internal packaging practices and the standardizing of package materials. (Package engineering has since become a nationally recognized profession.)

4. Installation of a scientific management program in the engineering shops, by hiring a team of machine tool experts to direct the work.

5. Securing the services of nationally known Carl Barth to design and construct slide rules for the company's metal working operations. Estimating time for shop operating standards was reduced from hours to minutes by Barth's slide rule adaptations.

Even the Depression of the early 1930s did not stop Armstrong from allocating a portion of its limited industrial engineering staff to research work in the field of management. Armstrong recognized the value of job evaluation as a management tool and was one of the first to establish a job evaluation plan. Its original plan, which was adopted in 1932, remained in effect until 1947 when it was revised and copyrighted. The Block Ranking Plan, which resulted from the original research, is still in effect and has been adopted by several other companies with Armstrong's approval.

In the pre-World War II years, other innovations appeared on the Armstrong industrial engineering scene. Group incentives were extended and a new concept of group bonus payments was developed. This development consisted of modifying the Bedaux concept of incentives for manual work to fit incentives for machine or process-paced operations. Armstrong's internally conceived Standard Performance Plan of the late 1930s resulted from this combination of circumstances.

During this time, another Armstrong first was recorded. Documented cost reduction programs became a regularly assigned industrial engineering responsibility. This was a planned program of profit improvement by item or project wherein the responsibility for the installation, administration, and follow-up (keeping score)—also by item or project—was performed according to a prepared schedule.

World War II brought with it a host of new industrial engineering activities in the company's munition manufacturing operations. Much of this expanded activity was brought about by Armstrong's entry into mass production metal fabrication operations. Industrial engineering services were extended to the production of shells, aircraft parts, and the manufacture of munitions. As a result of applying the theories of a processing industry to metalworking and assembly type operations, Armstrong became one of the nation's low cost producers of vital materials for World War II.

In the era immediately following World War II, Armstrong industrial engineers continued their pioneering role. Activities included:

1. Establishment of an industrial engineering center as a methods laboratory, model shop, and training headquarters. Ralph Barnes, an internationally known industrial engineering authority, was employed company to establish the facilities and conduct the initial activities.

2. Development of performance rating films for training both in dustrial engineering personnel and production supervisors in the concepts of work measurement. The procedure developed by Armstrong industrial engineers worked so well that the Society for the Advancement of Management adopted the Armstrong proce dure for producing their well-known and highly respected film series on a "Fair Day's Work."

3. Preparation of a company industrial engineering manual for use as an administrative guide in job evaluation and work measure ment. This industrial engineering publication also served as an in struction manual for training new industrial engineers and pro duction supervisors.

Concurrently, with the establishment of the industrial engineering center, company industrial engineers prepared a series of in-plant methods programs that were conducted company-wide for both plant supervisors and plant staff personnel. Subject material consisted of the basic techniques for analyzing factory operations such as flow charting, process charting, and man-machine charting.

Armstrong's pioneering methods have served as a model for many other companies.

Dow Chemical Company[2]

Toward the end of World War I, H. H. Dow, the founder of Dow Chemical Company, became interested in the new field of motion study techniques and principles. He obtained the services of one man to work along these lines. Unfortunately, this man's name and his accomplishments have been lost.

A group that later evolved into the present industrial engineering department at Dow Chemical was established in 1928. At that time, the Bedaux Company was hired to introduce a wage incentive plan. As the system was installed over the next five years a staff of stopwatch time study employees was established to obtain standards for incentive pay purposes. Typical of the degree of interest in this time study group, several top level employees were given assignments of several months duration in it. Included was Willard Dow, the founder's son, who later became a president of the company.

In 1933, the payroll department's functions pertaining to payment of bonus earnings were joined with the time study group under one supervisor as the standards and pay department. This name was eventually shortened to standards and the Bedaux group within it was referred to as time study. Until 1941, the major activities of the time study group were incentive-oriented. In 1941, a major union contract was signed and at the union's request all incentive plans were eliminated. Later the company and union agreed to reinstall incentives on certain highly undesirable jobs such as lime-unloading.

A government freeze on wage increases during World War II proved to be a boom for job evaluation work. The only way by which a pay raise could be enacted was to upgrade a job and then reevaluate it to increase the base rate of pay. Restructuring, job content, manning, and methods thus grew in importance, particularly where there was a chance to increase productivity.

During this same period, Dow industrial engineers working with an outside writer, developed the book, *So You Don't Like Time Study*, to help overcome some of the resistance to stopwatch studies. The book was so well received at Dow Chemical and elsewhere that tens of thousands of copies were sent upon request, to companies and educational institutions throughout the country.

The standards group led by Thomas Haines from 1940 to 1965 became more involved in the broader facets of industrial engineering,

[2]Material in this section is from information supplied by The Dow Chemical Company with permission of James G. Beeson, Manager of Industrial Engineering.

such as facilities layout, materials flow, and cost studies. The payroll and job evaluation functions were split off and the standards department was renamed the industrial engineering department. The department pioneered the use of operations research concepts in the Dow Chemical Midland Division before they had become an accepted tool in industry. It has also focused attention on cost reduction, methods improvement, and productivity studies in all operating areas.

Eastman Kodak Company[3]

In 1913, an industrial engineering division was organized within the Eastman Kodak Company at Rochester, New York. At first efforts were spent on measuring work and on installing and maintaining wage incentive plans for direct and indirect labor. Since then, the division has employed almost every industrial engineering technique known. Standards and incentive plans have been maintained through the years with a continuing history of change and improvement. Technological improvements since World War II and the recognition that employees are motivated by factors other than money led to the development of a new approach to the application of incentives. The resulting plan, titled the Individual Rate-Performance Premium Payment Plan, is described in the sixth edition of Ralph Barnes' textbook, *Motion and Time Study*.

During the first half of this century use of industrial engineering grew to the point that by the late 1960s individual industrial engineering departments were assigned to each of Kodak's major areas, including manufacturing, service, marketing, and product distribution. Each industrial engineering department was charged with the responsibility of providing the entire scope of industrial engineering services to its assigned area. (Figure 2 is an organization chart of Kodak's industrial engineering division as it existed in 1969.)

The auditing group randomly audited the work measurement and payment plans. The research and development group kept track of new and changing industrial engineering technology, acting as consultants to other groups on each of the major industrial engineering functions such as layout, materials handling, methods measurement, pay plans, queueing, linear programming simulations, computer applications and related software. The administration group was responsible for recruiting and placing personnel and administration of budgets, among other things. The special projects group was established. The human factors group was

[3]Reprinted, by permission of the pub lisher, from "Applying Industrial Engineering to Management Problems," *AMA Research Study #97*, ed. by A. W. Rathe & F. M. Gyrna, p. 50, © 1969 American Management Association, New York. All Rights Reserved.

FIGURE 2.
Organization of the Industrial Engineering Division at Eastman Kodak Company, Kodak Park Works.

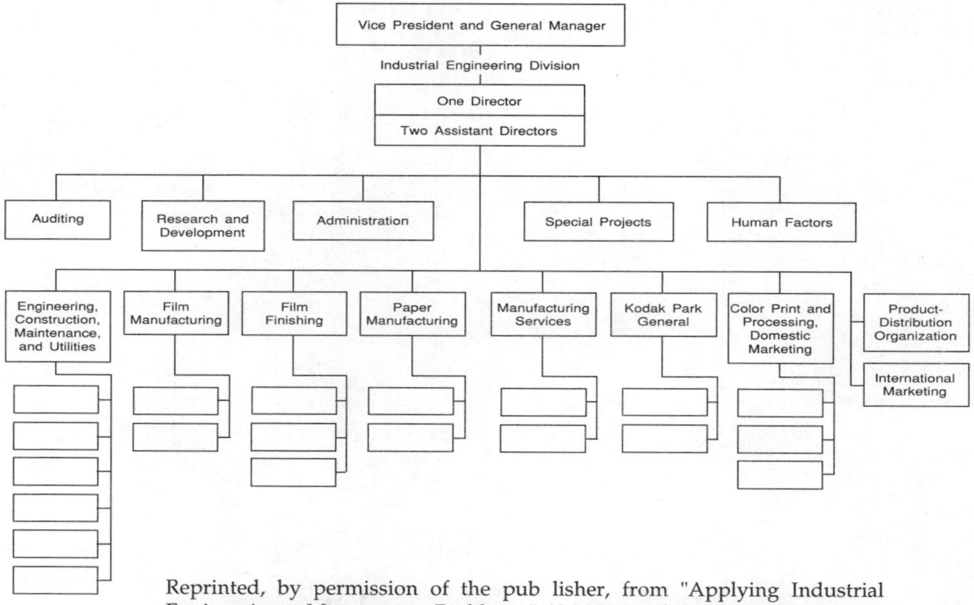

Reprinted, by permission of the pub lisher, from "Applying Industrial Engineering to Management Problems," *AMA Research Study #97*, ed. by A. W. Rathe & F. M. Gyrna, p. 50,

Table 6.
Industrial Engineering Functions and Services at Eastman Kodak Company, Kodak Park Works.

1. *Work Design*
   - a. Design of jobs
   - b. Operational methods
   - c. Man-machine relationships
   - d. Psycho-physiological studies
     - 1) Fatigue
     - 2) Work schedules
   - e. Design of workplace
     - 1) Facilities
     - 2) Convenience
     - 3) Fixtures
     - 4) Safety
   - f. Environments
     - 1) Temperature and humidity
     - 2) Noise
     - 3) Lighting
     - 4 Surroundings

2. *Operations Research*
   - a. Production system
   - b. Process analysis
   - c. Scheduling and planning work
   - d. Prediction models

3. *Work Organization*
   - a. Behavior patterns
   - b. Work simplification
   - c. Job enlargement

4. *Design of Production System*
   - a. Manufacturing alternatives
   - b. Organization
   - c. Control
     - 1) Physical
     - 2) System
     - 3) Waste

5. *Plant Location*
   - a. Space studies
     - 1) Current needs
     - 2) Long-range plant needs

6. *Plant Layout and Design*

7. *Materials Handling*

8. *Warehousing and Storage*

9. *Analyses of Equipment and Building Investments*

10. *Measurement of Work*
    - a. Standard costs
    - b. Performance standards
    - c. Premium-pay plans

11. *Quality Control Systems*
    - a. Sampling plant
    - b. Statistical control procedures

12. *Production Control Systems*
    - a. Labor planning
    - b. Capacity planning
    - c. Planning of services
    - d. Equipment utilization
    - e. Facilities utilization
    - f. Organizational studies
    - g. Inventory studies

13. *Value Engineering*
    - a. Material substitutions
    - b. Package engineering
    - c. Inspection procedures

14. *Customer Service*
    - a. Sanitation studies
    - b. Customer-order studies
    - c. Waiting-time studies

Specialized Services

1. Package Engineering Laboratory
   - a. Physical testing
   - b. Physical design
   - c. Field simulation
   - d. Package specification
2. Human Factors Laboratory*
   - a. Physiological limitations
   - b. Visual perception
   - c. Psycho-motor responses
3. Experimental Shop
   - a. Fixture development
   - b. Workplace design
   - c. Scale models

*Jointly administered by Industrial Engineering and Medical.

Reprinted, by permission of the publisher, from "Applying Industrial Engineering to Management Problems," *AMA Research Study #97*, ed. by A. W. Rathe and F. M. Gyrna, p. 51, a joint endeavor with the medical department. Included on the human factors team were engineers, psychologists, physiologists, and medical doctors. With work that began in 1959, an impressive human factors laboratory has been developed.

Eli Lilly and Company[4]

From 1913 through the early months of 1914, Eli Lilly and Company employed the consulting services of the Emerson Company (founded by Harrington Emerson). The Emerson engineers were known as efficiency experts in those days; today they might more properly be called industrial or management engineers.

The Emerson engineers conducted an exhaustive review of all operations at Eli Lilly and recommended the establishment of a bonus payment plan. This led to the formation of a methods and standards department in early 1914. During the next thirty years efforts were made to install wage incentive plans on all production operations and a few repetitive office jobs. During this same period, point-plan job evaluation was adopted.

After World War II, the range of industrial engineering activities widened. Methods and systems departments were organized in each operating division. The standards departments were combined with the methods and systems departments. Critical path scheduling, the use of breakeven charts on product introduction studies, and a variety of other industrial engineering techniques were adopted.

A corporate industrial engineering group focused attention on systems development and corporate-wide industrial engineering activities. This group introduced the concept of office landscaping to Eli Lilly and Company, a concept that employed the use of movable partitions and nonrectilinear layouts to use space effectively as well as decrease the cost of revising office spaces. Methods planning and facility planning functions were also added.

Stockham Valves and Fittings[5]

Progress in the application of scientific management at this Birmingham, Alabama, company took place step by step. During the period of 1920-1930, several hundred engineering standards were developed and imple-

[4]Material in this section is from information supplied by Joe C. Cook, Manager of Production Services, Elanco Products Company, a division of Eli Lilly and Company.

[5]Material in this section is from information supplied by Stockham Valves and Fittings with the permission of C. L. Pifer, Chief Industrial Engineer.

mented, which provided a factual and objective base for executive decision making as opposed to decision making by personal opinion and prejudiced preference. Included were design standards, pattern standards and equipment standards. They were the product of careful study and of the existing knowledge of the subject, and they served to condition management thinking .

The period 1930-1940 was devoted to the development of production standards. In 1929, a time study department had been established and the installation of a so-called modern incentive plan had begun. Since a prerequisite of wage incentives is methods study and standardization, this program paralleled the installation of the incentive plan and became an integral part of it.

The period 1940-1950 was marked by the installation of advanced personnel practices and techniques and resulting personnel standards. These extended into such fields as industrial health, industrial psychology, training, records, merit rating, employment standards, aptitude testing, safety, fire protection, and job evaluation.

Subsequent developments were the formation of a quality control department, which, in conjunction with a long-established inspection department, produced process controls to implement a defect prevention system. Sampling plans were accepted as an inspection technique. Many of the inspection functions were placed on a direct incentive plan. Also, participation with outside standards associations such as ASTM in formulating and complying with applicable standards emerged.

The Western Electric Company Company[6]

The Western Electric Company, a subsidiary of American Telephone and Telegraph, Inc., originated a number of areas of scientific management through innovative pioneering. In 1887, company foremen (as in other manufacturing companies in the United States and England) were, for all practical purposes, sub-contractors. Each hired his own staff of workmen, kept his own payroll, and promoted or discharged his men on his own responsibility. All shop employees except the foremen were hired and paid by the day. In some companies the foreman received a bonus for the completion of work before schedule. In 1896, this practice was discontinued at the Western Electric Company; the foremen were put on regular wages and hourly wage rates were established for the men.

[6]Material in this section is from information supplied by the Western Electric Company's public relations department.

In 1898, piece work was first introduced at the New York shop of Western Electric. In 1908, production teams, paid by wage incentives, were formed. Wage incentive rates were based on the concept of "a fair days work" and were usually expressed in average hours per unit of output for the whole team. (By 1951, more than 90 percent of the company's hourly rated employees who were directly engaged in manufacturing were a part of a production team. The concept of working together on a production team to relieve the monotony of repetitive jobs and enhance the enjoyment of making a living would have been appreciated by the Gilbreths.)

In 1924, Dr. Walter Shewhart introduced a new concept of economic quality control for manufactured products which he called "quality control charts." These charts applied statistics to the field of quality control.

In November 1924 the Western Electric Company, in cooperation with the National Research Council of the National Academy of Sciences, made plans to study the "relation of quality and quantity of illumination to efficiency in industry" at the Hawthorne Works near Chicago. These experiments lasted until April 1927, and had unexpected results in demonstrating that illumination was one among many factors that affect employee output. The human relations factors, about which little was scientifically known then, were much more important.

So, in 1927, a new series of experiments was started, this time in collaboration with Harvard University. The authoritative report on many of these experiments is presented in the book, *Management and the Worker* written by F. J. Roethlisburger of Harvard University and William J. Dickson for the Western Electric Company. Earlier studies on the effects of illumination were repeated. An experiment was made with rest periods., another with shorter working days and weeks, and another to test the effects of wage incentives. In all of these, the important factors seemed to be the attitudes of the workers' attitudes, including the fact that they were being studied. The amount of illumination, the timing of the rest periods, the length of the work day, and the amount of incentive wages were not as influential as were the human relations factors.

Such conclusions were not anticipated by Taylor or other engineers who had worked twenty or thirty years earlier. Taylor had experimented with rest periods, which he thought should be more frequent and longer for jobs involving heavy loads, and concluded that there was a definite relationship between rest periods and employee output. Barth called this "the law of heavy loads." Taylor had based his incentive plans on the idea of the economic man, that the worker was interested in the money and that the dollar incentive was the complete answer to attaining adequate productivity. The Hawthorne studies indicated that money was not the major objective of workers. In fact the authors of this study stated that, "None of the results gave the slightest substantiation to the theory that the worker is primarily motivated by economic interest" (Roethlisberger and Dickson 1939).

These studies at the Hawthorne Works of the Western Electric Company opened up a new outlook for the industrial engineer. They emphasized a need to learn more about the behavior of factory workers. Since the industrial engineer deals with people, as well as machines and processes, as a part of the systems which he analyzes or designs, courses in psychology and social sciences became important in the industrial engineering curriculum.

In early 1940, predetermined elemental time standards were developed based on electric time-recorder studies and the study of motion pictures of industrial operations. A system known as elemental time standard for basic manual work (ETS) was adopted for use in computing standards for wage incentive programs. An ETS manual enabled the industrial engineer to determine in advance of production how much time a job should take by identifying the motions required to perform the job and then summing the time values of each elementary movement.

Western Electric was a pioneer in the field of human factors which is now called biometrics. This is not human relations as was studied in the Hawthorne experiments, but a study of the latest data from the fields of physiology, psychology, and anthropology regarding human beings in relation to factory operations and their responses to signals and communications. Western Electric set up a manufacturing division task force on biometrics and a new engineering department to apply biometrics and biotechnology to manufacturing operations. Frederick Damon, a senior member of the AIIE and a member of the Human Factors Society, was placed in charge of the engineering department. Engineers cooperated with medical doctors in experiments on the design of tools and equipment. For a time, Western Electric secured the services of E. R. Tichauer, professor of industrial engineering at Texas Technological College. Shapes and sizes of the human anatomy were considered in redesigning pliers, chairs, work benches, screwdrivers, soldering irons, foot rests, and working environment, all directed toward improving employee comfort and well being so that he or she would perform better in the areas of productivity, quality, and reliability. Frank and Lillian Gilbreth would have been especially interested in this emphasis on the operator in relation to the task, the machine involved, and the design of the work place.

Curriculum changes resulted from industrial innovations in the new field. Engineering departments turned to other colleges for the life sciences and psychology courses, but added their own courses in human factors because of events that started in industry. The management movement and industrial engineering were characterized from the beginning by relationships between industry and education. Western Electric has cooperated with universities and technical institutions in the towns in which its plants are located to provide continuing education for its employees.

References

Roethlisberger, F. J. and W. J. Dickson. 1939. *Management and the Worker.* John Wiley & Sons, Inc. New York. pp. 575-576.

7. Consultants

The consultants who introduced the principles and practices of scientific management and industrial engineering to industry played an important role in the expansion of the new profession. Those who adhered to the concepts of Taylor and Gilbreth are to be commended. Some who introduced counterfeit techniques—parts of the scientific management system instead of the whole system—who took their fees and left without follow-up or training of successors, delayed the introduction of science to industry.

From the beginning, the consulting profession has maintained a close relationship with industry and academia. Some consultants had colorful personalities and language which impressed technical workers in industry. Most were good teachers, a good trait to have as a philosophy of management had to be transmitted as well as training in the new methods. Consultants to industry were in demand to lecture or give seminars in the universities.

The six functions of a consultant, as stated by Serge Birn, founder of the consulting firm by the same name, are:

1. To supplement the internal talents available in a company;

2. To offer counsel and advice, bringing in experience gained in other companies;

3. To diagnose problem areas: trouble-shooting;

4. To act as a catalyst, or go-between, for selling ideas to management;

5. To provide training and education; and

6. To innovate, combine, and transfer techniques, programs and solutions.

Perhaps the most important, though least recognized, function of a consultant is to bring the latest academic achievements to industry. An illustration is the Dartmouth College Conference on Scientific Management organized by Dartmouth's Amos Tuck School of Business Administration and Finance held in Hanover, New Hampshire in 1911. This conference, like many others since, provided an opportunity for representatives of management, universities, and the new industrial engineering profession to get together and exchange views. Many of the profession's early pioneers were at this meeting. Among them were Carl Barth, consulting engineer, Philadelphia; Morris Cooke, consulting engineer, Philadelphia; Harrington Emerson, president, The Emerson Consultants, Inc., New York City; Henry Gantt, consulting engineer, New York City; Frank Gilbreth, president, Frank B. Gilbreth, Inc., New York City; Lillian Gilbreth; H. K. Hathaway, vice president, The Tabor Manufacturing Company, Philadelphia, later a consultant; William Kent, consulting engineer, New York City; Frederick Taylor, consulting engineer, Philadelphia; Sanford Thompson, consulting engineer, Newton Highlands, Massachusetts.

Unfortunately, the proponents of scientific management failed to get the educators to realize the possibilities that the new field opened up to education. "Education," they were told, "is different from industry." This was the reply that the consultants heard from many other sources.

Lillian Gilbreth, the only woman to speak at the meeting, asked, "Isn't it merely a question of difference of vocabulary between academic work and scientific work? Why not bridge the gap and all go ahead together?" (Dartmouth College Conference on Scientific Management 1912). Frank Gilbreth reported that university graduates he had hired complained that their professors had not taken scientific management seriously and did not teach it while they were in school.

EARLY CONSULTANTS

If consulting engineers in the industrial engineering and management field can be grouped by generation, the first generation was composed of the pioneers of the field themselves and their immediate followers. The second generation was composed of those who adopted and refined the concepts of the pioneers.

Frederick Taylor

Taylor was retained as a management consulting engineer for Bethlehem Steel from 1899-1901. Prior to that he consulted for companies such as Simonds Rolling Machine Company, the Lorain Steel Company, and William Deering and Company. In 1903, he retired from professional practice to promote scientific management without charging fees. His followers continued to introduce management on a consulting basis.

Henry Gantt

Gantt became acquainted with the principles of scientific management through his association with Taylor at the Midvale and Bethlehem Steel Companies. In 1902, he opened an office as a consulting engineer. Included in his many clients were Brighton Mills, the Celluloid Company, Chevey Brothers, and Remington Typewriters. As a consultant, Gantt fostered the Taylor system adding ideas of his own, as other consultants did.

Carl Barth

Barth was a consultant at the Link-Belt Company, among many other companies, on introducing the Taylor system. It was his philosophy that companies should accept the entire system, not just part of it.

Frank Gilbreth

In 1913, the Gilbreths opened the first of four successive summer schools in the management field. By 1915, Frank had an extensive management consulting practice in Europe. In January 1925, after her husband's death, Mrs. Gilbreth started a motion study school to train students from industry. Other clinics and consulting contracts with industrial companies followed.

Harrington Emerson

In 1907, Emerson formed The Emerson Consultants, Inc. in association with his brother Samuel. Among the first clients was the Eli Lilly Company. Emerson Consultants was one of the organizers and a charter member of the Association of Consulting Management Engineers in 1933. Emerson was appointed a member of the Commission for the Elimination of Waste in Industry by President Herbert Hoover. He was also a member of the Civilian Expert Board on Industrial Management of the U.S. Navy, a board formed to review the functions and conditions of the navy yards.

LATER CONSULTANTS

The first generation of consultants in the field of industrial engineering and scientific management were the pioneers of the movement themselves and the immediate followers who worked with them in applying the new concepts in industry. A new movement might well die out after the first enthusiastic group moves out of the picture. In the case of industrial engineering, however, there developed a second generation of consultants who understood the principles and ideas of the founders and spent their lives applying them.

Charles Bedaux

Charles Bedaux, a French engineer, started his own consulting firm in 1917 in Grand Rapids, Michigan. The Bedaux Company moved to Cleveland, Ohio, in 1919; its first large client was the Robbins and Myers Company.

By 1920, its client list had expanded to include Remy Electric, a division of General Motors; General Electric; Brown Hoist; and Hood Rubber. The company survived the recession of 1920 and in 1921, work started at the Eastman Kodak Company in Rochester, New York. Offices were moved from Cleveland to New York and Boston. In 1926, offices were opened abroad in Italy, Germany, France, Persia, India, South America, and Australia. Southern operations were opened in 1927 at the Thomaston Cotton Company in Georgia. A 1928 client was the Dow Chemical Company.

During the Depression, Bedaux tried to bring the Duke and Duchess of Windsor (Edward VIII) to America for a grand tour. This troubled labor and management alike during this period of economic depression and led to the end of Bedaux as an American management consultant. The company's New York staff resigned to form a new company, American Associated Consultants.

Bedaux developed principles of industrial engineering that are still in use today. His invention of "B" units was used in incentive and evaluation systems. A man-minute or "B" covered a fraction of a minute for relaxation, with these fractions varying with the strain of the job. Actual "B's" could be compared with standard "B's". Periodic auditing of the standards was an important part of a system of control.

The principles which the company demanded that its clients follow are basic tenets of industrial engineering: pay a fair hourly wage based on the worth of the job for every hour spent in the plant regardless of production; base productive standards on time study analysis unimpeded by the opinion of foremen, accountants, or management; standards so set must remain unchanged as long as the process remains the same.

Phil Carroll

Phil Carroll built on and expanded the work of Taylor and Gilbreth. He felt that consultants should believe in and adhere to certain principles. The primary concept that guided him was what we now call "standard data," the idea that all jobs represent various combinations of the same elementary motions making it possible to set standards on complex jobs by adding each motion involved together.

From his start in consulting, Carroll was careful to separate indirect labor from direct labor. He operated on Taylor's indication that work done in the past by workmen should be divided in two parts with one part to be taken over by management. By dividing work into parts, errors could be found more readily. Carroll usually found errors in the part taken over by management. This was probably due to management's failure to plan, specify, and follow-up on projects.

Unlike Taylor and Gilbreth, who said that one should perfect a method before making a time study of it, Carroll studied the job as he found it. His philosophy was to make savings through improving productivity before spending money to make changes.

Carroll built his consulting life around one direct objective, that of improving time study fundamentals. He thought of time study as the basis for other engineering and management action. In correspondence with author Howard Emerson, Carroll wrote, "I am striving to get a correct foundation for inventory, part costs, overhead distribution and cost of sales. This notion will become increasingly important as we move into 'management systems'." His consulting work involved him in providing the data for such systems.

Carroll deserves credit for certain innovations, inventions, and improved procedures. He had an important part in the development of performance films. In December 1938, he gave a talk at a Society for the Advancement of Management (SAM) annual conference in which he said, "It is necessary to rate each time study in terms of some pace." As a result of further discussion on this matter and of Ralph Presgrave's paper on effort rating in October 1939, the suggestion was made that SAM form a rating committee. This committee developed the pioneer SAM films which were widely used in industries and universities. Carroll accumulated a library of 150 rating films.

Carroll patented a time study machine and was involved in the development of multiple and variable charting. The latter is described in his book, *How To Chart Timestudy Data*, published in 1950.

In addition to his contributions of serving as an officer and committeeman for many societies, and in publishing widely accepted factory textbooks, Carroll spread his industrial engineering know-how and management philosophy through presentations and seminars at plants, universities, and other organizations.

He lived in the same town with and was a close friend of Lillian Gilbreth. In 1969, he was awarded the Frank and Lillian Gilbreth Industrial Engineering Award from AIIE. He also received the Wallace Clark award from SAM in 1965 and the Gantt Medal of 1949-1953 from the ASME. A summary of his work and an expression of the respect that industrial engineers felt for Carroll are presented in the following excerpt from an obituary statement in the December 1971 issue of *Industrial Engineering*:

> A Fellow of AIIE, Phil was active in its local and national affairs, and served as a Director of the Work Measurement and Methods Engineering Division. In 1969 he received the Frank Gilbreth Industrial Engineering Award in recognition of his outstanding contributions to the profession.
>
> Other honors accorded Phil include Fellowship in the International Academy of Management and SAM's Gilbreth Medal and Taylor Key. He was a past president of SAM, and a member of ASME, Alpha Pi Mu, and Tau Beta Pi as well as AIIE. He held B.S. and M.S. degrees from the University of Michigan.
>
> Phil's strength as an advocate of sound principles will be a great loss to the IE profession.

Wallace Clark

Wallace Clark was a friend of the Gilbreths and a follower of Gantt. He built his consulting program around the Gantt Chart and is known for advocating scientific management in Europe. His book, *The Gantt Chart* (Clark 1922), was translated into many languages. After his death, his wife, Pearl, assumed management of his consulting organization. In 1952, she established the International Institute of Management.

In his consulting work, Clark emphasized one of Gantt's key principles: collect all the information you know about a job, improve it to the highest degree, and teach it to others so they can carry on while you develop something better.

H. B. Maynard

H. B. Maynard developed methods-time-measurement (MTM) in collaboration with G. J. Stegmerten and John Schwab. He founded his own consulting firm, which is still in operation today, and in 1934 he founded the Methods Engineering Council for research in industrial engineering in Pittsburgh, Pennsylvania. As a student at Cornell University, from which he received a degree in mechanical engineering, he was a leader in many professional organizations. Maynard was an author and co-author on the subject of work methods and measurement, and was the editor of the *Industrial Engineering Handbook* published by McGraw-Hill, Inc. (The first edition appeared in 1956).

Allan Mogensen

Allan "Mogy" Mogensen served as an assistant professor at the University of Rochester from 1924 to 1930 and gained recognition as a prominent industrial engineer in 1932 when his book, *Common Sense Applied to Motion and Time Study* was published. He was consulting editor of *Factory Management and Maintenance,* a magazine concerned with industrial engineering, from 1933 to 1937. In that position Mogensen was one of the first to advocate that motion and time study training be given to large numbers of a company's staff. By 1937, Mogensen had conducted several successful work simplification programs for well-known industrial organizations. In the summer of that year, he trained foremen and supervisors in industrial engineering techniques in the first of many annual work simplification conferences held in Lake Placid, New York.

The phrase, *work simplification,* had been suggested to Mogensen by Ervin Shell of the Massachusetts Institute of Technology in 1932. Mogensen and others such as Lillian Gilbreth and David Porter used the term to include the training of workers and supervisors in the basics of method study to make them better able to participate in the improvement of their jobs. While only four people attended the first Lake Placid session, the course's reputation continued to grow. Over the years, work simplification techniques were adopted and refined by hundreds of firms whose engineers and supervisors found them to be a positive way to improve profits as well as employee satisfaction through involvement and participation. Also well attended were his executive conferences held in Sea Island, Georgia, beginning in 1950.

During World War II, Mogensen served as an expert consultant to the secretary of war. In this capacity, he applied work simplification techniques in defense industries and in the armed services. Mogensen did not invent new techniques; rather he devised new and better methods for presenting techniques. He explained in simple terms what Taylor and Gilbreth had previously stated. This explanation enabled others in an organization to better understand and appreciate the work of the industrial engineer.

Marvin Mundel

Marvin Mundel was a professor of industrial engineering at Purdue University and carried on consulting work for many years. He also supervised the motion and time study laboratory at Purdue. In 1945 he originated and developed "memomotion study" and wrote about it in the second edition of his book, *Motion and Time Study: Principles and Practice* (Mundel 1955). This is a form of micromotion study in which motion pictures are taken at unusually slow speeds. The idea developed to aid Mundel in writing the article "Easier Housekeeping" for *Life* magazine (1946). Many successful applications of this technique have been found in industry and government.

In 1952, Mundel developed a management engineering program for the United States Army at Rock Island, Illinois. In the following year he left Purdue to form the management consulting firm of M. E. Mundel and Associates. He has authored many articles including the definitive article on industrial engineering in the 1963 edition of *Encyclopedia Britannica*. Mundel served as president of AIIE in 1979.

A. B. Segur

When A. B. Segur developed the first predetermined time system, motion-time-analysis (MTA), in 1924, he was carrying out the dream of both Taylor and Gilbreth: that standard time values could be developed for motions of work. A relationship between the fundamental motions that Gilbreth called therbligs and the times required for their performance had been lacking.

Segur accepted the general principles of the therblig classification and, after countless studies in his own laboratories and in many plants and offices, made a discovery that came to be known as Segur's Law: "Within practical limits, the times required of all expert workers to perform true fundamental motions are constant" (Alford 1928).

Segur developed an analysis method that could be applied to all classes of work. He set up a program to train others in the use of MTA. Segur's engineers made original studies and observations in a client's plant or office, suggested methods improvements, installed incentive plans, and trained company personnel in the use of MTA. The only equipment needed was paper, pencil, and a trained eye.

Four of Segur's clients were the Field Packing Company, the Link-Belt Company, the Colliers Corporation, and the Hood Rubber Company.

Professional Engineering Registration

Today's professional engineering license recognizes engineers who have met certain qualifications and passed a registration test in their area of expertise. Engineers have not always had this opportunity of recognition. In 1951, Nathan Dougherty wrote *Engineering Registration* which contains the history of engineering registration. Dougherty had helped guide registration and organization of engineers in the United States during a long career as an engineering educator and was a past president of the American Society of Professional Engineers.

For full protection, consulting engineers in all fields need licensing laws. Problems in surveying and the need to avoid overlapping of jurisdiction of water rights in Wyoming led that state to pass a registration law in 1907. In 1908, Louisiana passed a licensing law establishing the limits for registration of civil engineers. This was to protect the public from inferior structures as well as for the protection of the engineers themselves.

Registration of engineers did not immediately take hold. The American Society of Civil Engineers developed a model law and the American Association of Engineers sponsored legislation, but adoption was limited. Some engineers were afraid that registration would restrict the meaning of "engineer" only to those who had registered to practice.

After several states passed registration laws, a group of engineers from New York and New Jersey decided to organize a society patterned after the American Medical Association or the American Bar Association, in which all members would be licensed practitioners. Thus the National Society of Professional Engineers (NSPE) came into being in 1934. This society was formed to study salaries, working conditions, and public appreciation of the engineering profession. Today it has the task of moderating state and national laws relating to engineers. The NSPE began an active campaign for the extension of state laws. Technical engineering societies were lukewarm toward registration; some opposed it.

In 1920, as soon as states began setting up licensing laws and state boards for the registration of engineers, the National Council of State Boards of Engineering Examiners was formed. An objective of this organization was to improve uniformity in state examination requirements. Professional engineering licensure for industrial engineering is coming about state by state.

Engineers today are encouraged to join the technical society of their field as well as the NSPE, which can protect their legal rights. The technical society is concerned with developing the field of knowledge of the engineer, and is not permitted to lobby or support legal rights of the engineer. On the other hand, the NSPE with its state affiliations is in a position to protect engineers at a time when their projects are beginning to have greater social and political implications and relationships. Recently, the interests of the two types of societies have merged and agreements are being made between the NSPE and some of the national technical societies.

A number of organizations of firms and individuals representing consulting engineers have been organized, such as the Association of Consulting Management Engineers (ACME) headquartered in New York. (Wallace Clark was instrumental in founding this association.) Members of ACME are consulting firms; individual consultants join the Institute of Consulting Management Engineers, also located in New York.

References

Alford, L. 1928. *Laws of Management*. The Ronald Press Company. New York. p. 178.

Carroll, P. 1950. *How To Chart Timestudy Data*. McGraw-Hill, Inc. New York.

Clark, W. 1922. *The Gantt Chart: A Working Tool of Management*. The Ronald Press Company. New York.

Dartmouth College Conference on Scientific Management, October 1912. Dartmouth College. Hanover, New Hampshire.

Dougherty, N. 1951. *Engineering Registration*. The University of Tennessee Press. Knoxville, Tennessee.

Industrial Engineering. 1971. American Institute of Industrial Engineers. New York. 3(12): 10.

Mogensen, A. 1932. *Common Sense Applied To Motion and Time Study*. McGraw-Hill, Inc. New York.

Mundel, M. 1946. Easier housekeeping. *Life*. September. Time-Life, Inc. New York.

Mundel, M.E. 1955. *Motion and Time Study: Principles and Practice. Second edition*. Prentice-Hall, Inc. New York. p. 287.

8. Influences of Government, Law, and Labor

The paper, "The Shop Order System of Accounts," by Captain Henry Metcalfe was one of the first on centralized accounting of shop procedures (1886). Metcalfe had been superintendent of the Franklin Arsenal, the Benicia Arsenal, and at the time of the paper's publication, was superintendent of the Watervillet Arsenal of the United States Government in Troy, New York. Frederick Taylor participated in the discussion of this paper at the ASME conference. The paper, which followed Henry Towne's presentation of "The Engineer as Economist," referred to the importance of keeping records both as a precise statement of authorized work and as a means of assembling costs for services and materials. Metcalfe's system was based on the following two principles, (stated in his own words):

> 1. The radiating from a central source, let us say the office, of all authority for expenditure of labor or material. These being, however they may be disguised, the elemental forms of all internal expenditure.
> 2. The converging toward the office from all circumferential points of independent records of work done and expenses made by virtue of that authority (Metcalfe 1886).

Taylor had been developing a similar system at the Midvale Steel Company. An advantage of this system, according to Taylor, was that a foreman would be able, within minutes' notice, to place his hands on the record of a piece of completed work before he started to make the same piece again for a subsequent order. Although orders would originate from a central office, records could be kept on cards by the foremen as well so that they would be readily available. The records would include information such as mistakes made on the job, method suggestions, cost, time involved, the employee who did the work, etc. The ideas expressed in this and other papers to follow led to the setting up of production control systems in public and private manufacturing plants.

In the 25th anniversary edition of *Industrial Engineering*, William Foster, a director with the United States Postal Service, wrote:

As early as 1906, key government personnel were considering principles of scientific management by Frederick Taylor for use at industrial facilities and test implementations were being conducted at Rock Island Arsenal, Mare Island and Boston Shipyard. In 1909, the Ordnance Department of the U. S. Army adopted the Taylor system for use in the manufacturing arsenals. By 1918, over one-third of the members of Taylor's Society for the Promotion of the Science of Management were working for the Federal Government in the Ordnance Department. Shortly thereafter, unfavorable reaction to the concept of scientific management developed in both industry and government, culminating in legislative prohibition against so-called 'efficiency experts' conducting time studies in government activities (Foster 1973).

Union Attitudes

Unions have historically opposed the use of stopwatch time study, wage incentive systems, and job evaluation. This opposition, they said, was based on the abitrary and abusive use of these methods and because results were often inequitable and not objective. Yet many companies today use time study and wage incentive systems with union agreement. An understanding of the system's merits by all is essential.

Many industrial engineers felt that unions had less objection to new techniques of measurement such as predetermined motion-time systems, work sampling, statistical techniques, and electronic timing and control devices than to direct stopwatch studies that cause the engineer to stand over the worker with a stopwatch board. Yet at the fourth Constitutional Convention of AFL-CIO on December 12, 1961, a resolution on industrial engineering pointed out that these new techniques were nothing more than subtle forms of the old.

In 1913, Congress stipulated that no part of a Navy appropriation bill should be made available for the pay of any person engaged in time study work. At that time, employers had misused time study standards by cutting piece rates after they had been set; so called "efficiency men" had often set rates rather carelessly.

One of the first adverse reactions to Taylor's system of scientific management came from the Norfolk Naval Shipyard. It was Taylor's intention that labor and management would join in the common objective of sharing their profits, but misunderstandings developed early on the part of management and labor as to what Taylor had in mind. An engineer who knew of Taylor's work attempted to introduce this method at the Norfolk Naval Shipyard but ran into opposition from the shipyard employees. The union opposed the use of stopwatches for time study and threatened anybody who brought a stopwatch in the yard with being dumped into the Chesapeake Bay.

While working one summer with the industrial engineering department at the shipyard, author Howard Emerson met the industrial engineer who first tried to bring a stopwatch into the shipyard. In an interview, he said the workers did not carry out their threat to throw him into the bay, but did throw hot rivets at him. He took his stopwatch home and did not bring it back.

In 1950, a consulting firm recommended that the Navy avoid this opposition by approaching the application of scientific management in a different way, emphasizing production control, specifically the scheduling of work. Work schedules would, of course, involve time standards. Typical problems that the shipyard was experiencing were having too much or too little time for repair work while the ship was berthed at the shipyard. A better production schedule would produce better repair work assignments, making the shipyard more efficient and therefore bringing the shipyard more business. Since the competition between shipyards for repair work was intense, any edge over the competition was valuable. With proper control of production, an engineer could inspect the repair needs of the ship as it came into the bay and make out a work schedule that could reasonably be completed while the ship was in port. These schedules had to be based on better standards than any available in the Norfolk shipyard, which led to a new set of studies by work analysts trained by university personnel. Some of the work standards developed were based on past records; some were engineered standards.

Before the program could be implemented, the shop superintendents had to approve it. Since high level supervisors had been trained at the Wharton School of the University of Pennsylvania and were acquainted with process charting and the procedures of measuring work, they did not object to this approach to the introduction of industrial engineering.

Defense Departments

In 1949 there was a movement to continue the restrictions of the 1913 Naval appropriation bill to the rest of the U.S. Defense Department including the Army and Air Force, both of which were using time study and methods improvement to their advantage. Senator Salonstall of Massachusetts tried to persuade Congress to approve the prohibition of stopwatches even in the departments that still wished to use them. His action was opposed on the Senate floor by Senator Russell Long of Louisiana and Senator Ralph Flanders of Vermont, who owned machine shops. (The discussion in the Senate on National Military Establishment Appropriations can be found in the *Congressional Record* for 1949, 81st Congress, 1st session, pages 12,318-12,323).

In the discussion, Flanders made several points that are applicable today: 1) time study has been the means by which the standard of living in this country has been raised; 2) the worker does not work harder, he works easier; 3) there is no reason for ruling it out of government works anywhere or at anytime; 4) some unions insist that time study and standards be included in union-management contracts.

As a result of this discussion, the ban against stopwatch use was dropped from the appropriation bill.

A Change of Position

In Foster's article, he noted that:

> The modern impetus of industrial engineering in the Federal Government began in 1948, when President Truman commissioned Herbert Hoover to conduct studies of the effectiveness and efficiency of government activities and to formulate recommendations of improvements of management practices. These studies became known as the "Hoover Commission Study" and were the basis for the President to formally issue an executive order in 1949 requiring that management improvement programs (industrial engineering) be initiated by all department and agency heads. In addition, recommendations by the commission eventually affected the lifting of the restriction on the use of mechanical timing devices (Foster 1973).

The United States Postal Service

In 1960, steps to introduce industrial engineering in the postal service were discussed by Senator Estes Kefauver of Tennessee. The organization of the postal service as a private-public corporation aided in making it possible to introduce industrial engineering concepts similar to that in private companies. However, in the same year, Kefauver introduced Senate Bill S3132 which prohibited the use of stopwatches or other time measuring devices in the postal service. Fourteen similar bills were introduced in the House of Representatives. This type of activity had been going on for a long time, although over the years, none of these bills got further than the Committee of Post Offices and Civil Services. Presumably, however, those who introduced the bills got credit from labor unions for sympathetic action.

The fact that a congressman with Kefauver's reputation had introduced a bill in the Senate suggested that there was a greater than usual danger that the basics of industrial engineering would be kept out of public industries. In July 1960, Kefauver was informed of industrial engineers' concerns in this matter. A letter expressing this concern was written to him by the summer 1960 industrial engineering class at the University of Tennessee. Kefauver did not introduce any more bills of this kind, however, they continued to appear.

While there is no formal agreement as to stopwatch use in the United States Postal Service, there is an informal understanding that stopwatches as such will not be used. This does not preclude the use of MTM or similar predetermined time standards. G. B. Bailey, who represented the Canadian firm of J. D. Woods and Gordon, Inc., under a contract to introduce time standards in the post office (see *The Journal of Industrial Engineering*, May-June 1960, pp. 247-250) used the Basic Motion Timestudy (BMT) predetermined time system. Even the simple problem of deciding whether a mail delivery truck should go down one side of a street and back the other, or criss-cross, requires some type of time measurement. Without time measurements, the industrial engineer cannot make meaningful contributions.

Industrial Engineering and the Courts

Industrial engineers are called upon from time to time to give testimony in courts for—and against—labor unions, governments, etc. In the controversy between the General Electric Company's Hickory, North Carolina, plant and the National Labor Relations Board, four issues were at stake.

1. Refusal to permit an industrial engineer brought in by the union to inspect the predetermined motion-time data used by General Electric.

2. Refusal to permit the union's industrial engineer to make his own stopwatch time studies on the jobs in question.

3. Refusal to permit local union officers or the union's industrial engineer to observe and independently evaluate the jobs.

4. Refusal to supply the union with an area wage survey containing wage rates within each surveyed company.

Bertram Gottlieb, a professional industrial engineer and member of AIIE, appeared as an expert witness for the union. He pointed out that he could not judge the fairness of the standard without access to the company's motion time study (MTS) data and an opportunity to view the job and make his own time study.

A trial examiner ruled in favor of the union on two of the issues and in favor of the company on the other two; however, the National Labor Relations Board, upon reviewing the trial examiner's findings, reversed some of his rulings and ruled in favor of the union on all four issues. General Electric refused to accept the board's order and petitioned the court to review and set aside the board's decision. Clement Haynsworth, Chief Judge of the United States Court of Appeals for the Fourth Circuit in Richmond, Virginia, reviewed the decision and upheld it on all counts. (Haynsworth was later recommended but not approved to fill a Supreme Court vacancy.) General Electric appealed the Circuit Court's decision to the U. S. Supreme Court. (See *Industrial Engineering*, October 1969, pages 54-55.)

In another case that reached the U. S. Supreme Court, the court ruled that unions may penalize members who exceed set production quotas, under certain circumstances. The case was *Russell Scofield et al., employees of the Wisconsin Motor Corporation versus the National Labor Relations Board*. In the opinion written by Supreme Court Justice Byron White, it was noted that in 1938 the union initiated a production ceiling. Its members could accept immediate piece work pay for work above a standard but not beyond the ceiling. In 1944, this was made a formal rule enforceable by union fines and expulsion; one dollar for each violation and one hundred dollars for repeated violations. The margin between the so-called machine rate, a standard set by contract, and the ceiling rate set by the union was ten cents per hour in 1944, but as a result of collective bargaining over both the machine and ceiling rates, the margin was increased to between forty-five and fifty cents. The company regularly urged the union to abandon the ceiling and never agreed to refuse employees immediate pay for work done over the ceiling. Also, there was a mechanism for banking work done over the ceiling to be paid for in later slack weeks so that the production ceiling would not be exceeded. The petitioners, however, wanted and accepted immediate pay for work over the ceiling for which they were subsequently fined by the union. They appealed to the National Labor Relations Board, were turned down, and took the case to the Supreme Court.

Industrial engineers should be concerned by comments made by White in his review, which read in part:

> As both the trial examiner and the court of appeals noted, union opposition to unlimited piece work pay systems is historic. Union apprehension, not without foundation, is that such systems would drive up employee productivity and in turn create pressures to lower the piece work rate so that at the new higher level of output employees are earning little more than they did before. . . .
>
> Petitioner purports to characterize the union rule as featherbedding, but it is hard to square this with his collective bargaining agreement that an average, efficient employee produces at a "machine rate" substantially below the ceiling.

The Supreme Court ruled that the union ceiling left the bargaining process unimpaired, breached no collective contract, required no pay for unperformed services, induced no discrimination by the employer against employees, and represented no dereliction of its duty of fair representation. It affirmed that the union rule was proper.

In dissent, Supreme Court Justice Hugo Black held that the union coerced employees through fines for refusing to engage in a concerted activity. He referred to his dissenting opinion in *National Labor Relations Board versus Allis Chalmers*, 388 U. S. 175 199 (1967).

The Industrial Engineer and Arbitration

Industrial engineers have played, and continue to play, many roles in the labor arbitration process, from providing expert testimony during arbitration hearings to actually being arbitrators. Usually, the industrial engineer is in the position of testifying to an arbitrator who is not an engineer.

Paul Lehoczky, for years an arbitrator and head of the industrial engineering department at The Ohio State University, once pointed out that

> Of some three hundred members in the National Academy of Arbitrators and the thousands of individuals listed with the American Arbitration Association and the Federal Mediation and Conciliation Service as being available to act as arbitrators in industrial labor-management disputes, only a very small fraction, probably less than 1% are indus-

> trial engineers. Yet grievances are constantly being raised in those areas of industrial activity which are of vital concern to industrial engineers. Although most of these issues are being arbitrated by conscientious and capable arbitrators, many of these arbitrators do not possess the background in industrial engineering needed for the full comprehension of the effects of their decisions upon the operation of the plant. In consequence, their decisions will be heavily affected by how the case is presented and by the specific language of the agreement between the parties then in existence (Lehoczky 1966).

One example of an industrial engineering type of dispute involved a company in Tennessee that had a union contract. The contract provided for two types of grievances: one covering personnel and working condition grievances, and the other pertaining to industrial engineering types of grievances such as those related to work standards and job evaluation. For the latter type of grievance, Nathan Dougherty, the dean of engineering of the University of Tennessee, was asked to choose an industrial engineer as an arbitrator.

References

Foster, W. 1973. Government. *Industrial Engineering*. American Institute of Industrial Engineers. Norcross, Georgia. 5(9): 24.

Lehoczky, P. 1966. Industrial engineering and collective bargaining. *Labor Law Journal*. July. Commerce Clearing House, Inc. Chicago, Illinois. p. 393.

Metcalfe, H. 1886. The shop order system of accounting. *Transactions of the American Society of Mechanical Engineers*. American Society of Mechanical Engineers. New York.

Industrial Engineering. 1969. Industrial engineering and the Supreme Court. October. American Institute of Industrial Engineers. New York. 1(10): 54-55.

9. A Profession Matures

The history of civil engineering can be told by describing the structures, highways, and canals that civil engineers have built over the years. The history of mechanical engineering often centers on machinery such as the steam engine and the automatic machine tool. Chemical engineers identify with the development of new processes. The industrial engineer can point to the factory as the starting point and center of development of his or her profession. Although civil engineers and architects may have helped with the structure itself, the mechanical engineer with the design of the machinery inside, and the chemical engineer with the chemical processes, the industrial engineer became involved with the organization and design of the work place and the flow of materials through the factory. Today's industrial engineer is concerned with a variety of systems as an entity.

The idea that a profession with such diverse interests as are discussed below can be thought of as a "community" was developed by William Pollard. Pollard is retired from a career as a nuclear physicist at Oak Ridge, Tennessee. He is also a priest. His thought of a profession as a community sprang from the fact that a profession is made up of people who understand one another, "speak the same language," have the same apprach to problems, belong to the same organizations, read the same publications, and are trained in similar educational curricula. This brings about a community of interest.

As the original concepts of industrial engineering were refined, new technology was developed and research uncovered newer concepts. The profession matured to the point where it was to become accepted in service industries such as hospitals, retail stores, and banking. Those who were involved in this process and their contributions will be discussed in this chapter.

WORK DESIGN AND MEASUREMENT

Work design and measurement is a basic technique behind industrial engineering. It involves measuring work so that standards, or expectations of work to be completed in a certain amount of time, can be determined.

Then using this data, daily work schedules can be designed from production schedules. All areas of industrial engineering, such as plant layout and material handling, focus on doing work efficently. But work design and measurement is directly concerned with the actual work to be completed.

As described earlier, Frederick Taylor originated the idea of using time values to design jobs. He analyzed the elements of work, eliminated or improved some of them, and, by using a stopwatch, determined the amount of time required by each element. Frank Gilbreth identified basic motions common to most manual work, which he named therbligs. From this research, Gilbreth devised rules for motion economy and efficiency and published them in the *Society of Industrial Engineering Bulletin* (1923). Others added to this list of rules including H. F. Krueger, a Segur engineer, who identified fifty-one different principles of motion economy—improvement principles associated with individual motions. The presently accepted list of these principles was developed by Ralph Barnes and has appeared in his text, *Motion and Time Study*, since 1937.

The Westinghouse Electric Corporation has been a leader and pioneer in the application and use of traditional motion and time studies. A four factor rating system was published in 1927 and, as originally conceived, included skill, effort, working conditions, and consistency factors, each having a scaled quantity from low to high and each having an interaction with the other factors (Lowry, Maynard, Stegmerten 1940). Using the values arrived at from this system a time study engineer could obtain an objective evaluation of an operator's performance level.

Work sampling was developed in England by mathematician and statistician L. H. C. Tippett. He found it difficult to use stopwatch studies to effectively determine and measure causes of stoppages and worker delays on more than eight textile looms at a time. He developed a method called the "snap-reading method" to quickly collect such data in a large mill. R. L. Morrow was one of the first to use Tippett's method in this country.

Tippett's 1934 paper, "Statistical Methods in Textile Research. Uses of Binomial and Poisson Distributions: A Snap-Reading Method of Making Time Studies of Machines and Operatives in Factory Surveys," appeared in the November issue of *Shirley Institute Memoirs* and has been reproduced in the 1956 and subsequent editions of Barnes' *Work Sampling*.

Industrial engineers who are involved in work measurement often play a role in the administration of hourly rates. This role may vary from advising the personnel department about specifications and descriptions to managing a job evaluation program. The need for a sound wage administration program was accentuated in 1942 by a wage freeze caused by the Economic Stabilization Act. Only through a formal job evaluation program could an employer identify and correct an errant wage. The popular textbook, *Job Evaluation* (1948), was co-authored by Jay Otis, psychology professor at Case Western Reserve University, and Richard Leukart, the assistant secretary of the National Screw and Manufacturing Company. Industrial Engineers such as H. G. Thuesen and H. Lohmann also contributed to developments in job evaluation.

Predetermined Time Systems

Both Taylor and the Gilbreths had contemplated a list of standard times for each element in every trade. Bertrand Thompson, in conjunction with Taylor, attempted to establish a series of element times applicable to concrete construction, but it was Gilbreth's concept of universal elementary motions—therbligs—that supplied the factor necessary to give substance to Taylor's idea. If, as Gilbreth had pointed out, the same elementary motions were used in various combinations and sequences throughout industry, then it was possible to establish standard times for these elements.

The first successful attempt to establish such time standards and to develop a well-integrated procedure for work measurement based on predetermined motion times was made by A. B. Segur, a consulting engineer in Oak Park, Illinois. The results of Segur's research, which began in 1919 and reached its final state of development about 1924, were made available only to his clients. Thus his system of motion time analysis (MTA) was not available to a large audience and its applications were limited.

Later in the 1920s, another system of predetermined times was developed in the telephone assembly operations of the Western Electric Company. This system, known as elemental time standards (ETS), along with Segur's MTA system provided the groundwork for present-day systems of predetermined times.

In the next two decades, many individuals and companies developed a variety of predetermined time systems. W. G. Holmes wrote the book *Applied Time and Motion Study* (1938) in which he described a "time of movements chart." This chart described tabulated values based on the measurement of body members. In the middle 1930s, Harold Engstrom of the General Electric Company constructed a set of data that he called Engstrom's Tables.

One system, motion-time standards, evolved from studies of electronic assembly operations of the Philco Company. Out of this work came one of the major predetermined time systems in use today, work-factor. A group of industrial engineers led by J. H. Quick at the Radio Corporation of America began to develop this system about 1934. It was not until 1945 that an article authored by Quick, W. J. Shea, and R. E. Koehler entitled "Motion Time Standards" appeared in *Factory Management and Maintenance*. This article gave a partial description of the procedure; it did not contain all the rules and tables but served to inform industry of its existence.

In 1940, the Westinghouse Electric Corporation engaged H. B. Maynard to conduct a study of time values for sensitive drill press work. This eventually developed into what is now known as methods time measurement (MTM). After several years of careful testing, this procedure was described in the book, *Methods Time Measurement*, (Maynard, Stegmerten, Schwab 1948). Maynard had the foresight to realize that research by one group was inadequate if confidence was to be built in predetermined time systems. MTM was not only made available to the public; in 1951, the MTM Association for Standards and Research was formed as an independent body to conduct research, maintain high standards of application, and provide members with pertinent information.

About 1950 a predetermined time system known as basic motion timestudy (BMT) was developed by the Canadian consulting firm of J. A. Woods and Gordon Ltd. The BMT system was built by analysis of factory operations and by laboratory experiments.

PLANT LOCATION AND LAYOUT, AND MATERIAL HANDLING

Industrial engineers often are called upon to work on plant location projects for private companies seeking branch locations, or for consulting firms, or for communities seeking new industries. Plant location involves collecting, compiling, and evaluating data necessary to make a decision for the best location of a plant.

In 1826, Johann Heinrich von Thunen wrote about a theoretical problem of plant location involving agricultural production and a central urban market. He postulated that the production location should be determined by the relationship between the product's market price and the distance from the market place. Thus, he considered only two of the many factors involved in a plant location decision.

In 1909, another German, Alfred Weber, wrote *Uber den Standort der Industrien* (1929) translated as *Theory of the Location of Industries*. He studied labor costs in addition to transportation costs and included a factor called the agglomerating factor. This is a "cheapening" of production caused by production being done at one place. Weber concluded that since transportation cost is a major factor, the production location should be at the source of the material if materials lose weight in conversion to product. Conversely, if weight is gained in product manufacture, the location of the production facility should be nearer the market.

The latest developments in plant location involve applying linear programming techniques to the problem of plant location. However, these techniques still require engineering judgment to correctly weigh tangible and intangible factors that are applicable to each individual case.

Plant layout (facility planning) is defined as "the field of selecting the most effective arrangement of physical facilities to allow the greatest efficiency in the combination of resources to produce a product or service" (Tompkins 1982). There are systematic techniques which have been developed in the curricula of industrial engineering and practiced with modifications in individual industrial companies.

As mentioned in Chapter 4, James Watt, Jr. and Matthew Boulton planned machine layout in terms of work flow requirements for their foundry in Great Britain that began operation in 1796. There, they erected various types of swinging cranes, an overhead traveling bridge crane and other lifting devices. About 1798, Eli Whitney planned the layout of equipment in his musket factory. In those days, material handling, closely associated with plant layout, consisted of handing parts of the musket from one operator to another.

The plant layout techniques that now are associated with high-speed production came into being in the early 1900s with the advent of the automobile. While Taylor's approach to increasing productivity was through motion and time study and the use of a wage incentive, the approach of Henry Ford's engineers was to eliminate labor by mechanical means. In August 1913, the first attempt at line assembly of an automobile was successful in increasing productivity. In the 1920s, other companies changed to conveyorized types of operation. Most of the basic types of material handling equipment were used during this period such as: platform skids and pallets, handlift trucks, powered platform trucks, and powered conveyors. However, not until World War II did the effective handling of material get widespread recognition by management as a means of increasing production.

Walter Rautenstrauch provided the first academic instruction in plant layout in a college curriculum in his work management course at Columbia University in 1907. Subject matter included the layout of mills and the character of construction with regard to lighting, ventilation, sanitation, insurance, and reduction of fire hazards. The first course in a regular industrial engineering curriculum was facility planning and design offered by Hugo Diemer at Pennsylvania State College in 1908.

In a study of industrial engineering practices published in 1956 by the National Industrial Conference Board it was noted that:

> Plant-layout activities are rarely an exclusive function of industrial engineering departments. While industrial engineering is frequently responsible for maintaining plant-layout drawings and information, the thinking behind the layout itself is generally the result of the combined consideration of many interested groups.
>
> However, because plant layout ties in so often with methods and the flow of work through the factory, industrial engineering departments often take a leading role. Some companies use committees, consisting of supervisors from industrial engineering, production and plant engineering, to coordinate plant-layout work.

ENGINEERING ECONOMY

Historically, the concept of depreciation appeared as far back as A.D. 27 in the writings of Vitruvius, where a figure for *n* equal to eighty years is mentioned in the valuation of buildings. The origin of interest calculations goes back to the days of Babylon, 2000 B. C. Although specialization of labor was practiced in armor factories in Rome and Greece, attention was first given to production economics by the great economist Adam Smith in 1776 at the time of the emerging factory system. In *An Inquiry into the Nature and Causes of the Wealth of Nations,* he observed three basic economic advantages resulting from the division of labor:

> This great increase of the quantity of work, which, in consequence of the division of labour, the same number of people are capable of performing, is owing to three different circumstances; first, to the increase of dexterity of every particular workman; secondly, to the saving of the time which is commonly lost in passing from one species of work to another; and lastly, to the invention of a great number of machines which facilitate and abridge labour, and enable one man to do the work of many (Smith 1976).

Adam Smith did not deduce these ideas in a theoretical manner. He observed these conditions of the factory system, noted the three advantages and recorded them. His book was a milestone in the development of production economics not only because of Smith's observations, probably accelerating the division of labor, but also because a scholar had recognized that there existed a rationale for production.

In 1877, Arthur Wellington wrote what is considered the first book on engineering economy, *The Economic Theory of the Location of Railways*. He used interest tables and the capitalized cost method in evaluating the design and location of railroad structures. The intent of his book was to be a guide for practicing engineers.

Rautenstrauch introduced what is now considered the first course in engineering economy, works management, at Columbia University in 1907. Courses in engineering economy have been offered in most industrial engineering curricula since 1931. The American Society for Engineering Education and the AIIE each established an engineering economy division.

In 1915, John Charles Lounsbury Fish wrote *Engineering Economics*. Fish, a civil engineer, believed that the central concern of engineering economy was the choice of investment, rather than of structure such as a building, road, or plant. He looked to Wall Street for his model and drew upon financial and actuarial mathematics for the formulation of his investment model.

O. B. Goldman in his 1920 book, *Financial Engineering*, set out to provide a definite scientific method of determining the comparative value of all things that the engineer must consider. He leaned heavily on compound interest calculations. His methods were limited in that certainty was the expected condition.

Eugene Grant's book *Principles of Engineering Economy*, published in 1930, followed along the general lines of Fish's reasoning. However, Grant realized that engineering economy was not synonomous with a process such as investing in bonds. In the book's first chapter, he defined an economy study "...as a comparison between alternatives in which the differences between alternatives are expressed as far as practicable in money terms." Grant also devoted a section of his book to judgment factors. Revisions followed, as did other books such as *Basic Accounting and Cost Accounting* (Grant 1956) which led the way in classical engineering economy. The American Society for Engineering Education (ASEE) presents an award named after Grant.

Benefit-cost analysis is a specialized area of study within engineering economy which became popular in the 1930s. Its purpose is to help government agencies determine the size of their budgets and which projects should be undertaken, based on the ratio of benefits to cost. If the benefit to cost ratio is greater than one, a project can be considered justified; the higher the ratio the more favorably a project is judged. A standard reference book on the subject was *Economics of Welfare* by A. C. Pigou in 1932.

The area of benefit-cost justification is also of great importance to engineers who work in public organizations and are responsible for public works that do not bring in income. A good example of this is the Tennessee Valley Authority (TVA). The generators and electricity producing areas of the TVA operation are income producing, but the locks and flood control barriers are not. Allocating investment costs and calculating benefits in such multi-purpose projects is a problem. Social costs and ecology deterioration caused by the projects must also be considered. New challenges in the field keep teachers and users of benefit-cost justification searching for new ways of solving problems.

PRODUCTION PLANNING AND INVENTORY CONTROL

Production planning is defined as "The function of setting the overall level of manufacturing output "to establish production rates that will achieve the set objective and keep the production force stable" (*Industrial Engineering Terminology* 1983). Inventory control is defined in the same book as "The activities and techniques of maintaining the stock of items at desired levels. . ." Early work in this area includes George Babcock's work at the Franklin Manufacturing Company in 1912 that attempted to solve the economic lot size problem. Babcock approached the problem on a mathematical basis by using a cubic equation. Federick Harris published one of the first economic lot size equations in 1915. His equation arrived at a minimum sum of inventory carrying costs and set-up costs where the demand was known and constant and the rate of production was assumed to be infinitely higher than the demand. Harris emphasized the necessity for ensuring that the assumptions on which the economic lot size formula was based were applicable, limiting the application of the formula. Harris' equation was modified in later years by other industrial engineers, including E. W. Taft (1918), Benjamin Cooper (1926), and Erich Schneider (1938).

Henry Gantt's chart took some of the first steps toward the systematic planning and control of operations. The first known use of the chart was in the artillery ammunition shops at the Frankford Arsenal when World War I was declared in 1914. The Gantt chart graphically depicted actual versus estimated production time. The principles of production planning and control established by Gantt are still in use, though they have grown more complex with the advent of the computer, predetermined time systems, critical path scheduling, and systems engineering.

Fairfield Raymond's book, *Quality and Economy in Manufacture* (1931), traces the historical development of contributions in the field of production and inventory control prior to 1930. Thomson Whitin's book *The Theory of Inventory Management* reviews systems in the field through 1951.

STATISTICAL QUALITY CONTROL

The quality of workmanship has been a concern throughout the centuries. Reliefs in a tomb in Thebes (1450 B.C.) depict inspectors checking the quality of stone blocks. In 1654, Pascal became interested in the probability studies related to games of chance. Together with Pierre Fermat, he formulated laws launching probability as an important branch of mathematics.

It was not until the 1920s that quality control and statistics were joined into statistical quality control (SQC). When the *Bell System Technical Journal* appeared in 1922, the early issues contained numerous papers on inspection and quality control. In 1924, Walter Shewhart of Bell Telephone Laboratories developed the quality control chart, presenting industry with a tool for evaluating variations in product quality. Shewhart's control charts were applied within the Bell Telephone Company and were supplemented by statistical sampling techniques developed by H. F. Dodge and H. G. Romig. Their article, "Sampling Inspection Tables" appeared in the October 1929 issue of the *Bell System Technical Journal*. Shewhart's *Economic Control of Quality of Manufactured Products* was published in 1931.

Use of statistical quality control began on a large scale in Britain before American industry began to reap the new technique's benefits. During a trip to Britain in May 1932, to speak before the Royal Statistical Society, Shewhart converted large segments of British industry to the use of quality control techniques.

However, the acceptance of statistical quality control in the United States--outside of the Bell Telephone Company--was slow until after World War II. Training courses were started in this new field at Stanford University in 1942. Organizers were W. Edwards Deming, Eugene Grant, and Holbrook Working. The first course was a ten-day course given to representatives of war industries, procurement agencies, and armed services. This and subsequent courses were part of the engineering, science, and management program sponsored by the United States Office of Education.

In 1943, the War Production Board created a team made up of Working from Stanford University, Olds from Carnegie Institute of Technology, and Paul Peach from North Carolina State College to conduct training courses in statistical quality control and to assist factory managers in its use. The training courses ended after the war but the nearly two thousand people who had participated in them became the nucleus of the American Society for Quality Control (ASQC).

Industrial engineering departments began incorporating statistical quality control in their curricula and many collaborated with ASQC chapters in spreading this new concept, including a technique through which the use of statistical limits could detect potential defects before specifications were exceeded. Meanwhile, American management and workers looked aside, seemingly lacking in the necessary mathematical background and desire to put quality first.

The first textbook on SQC, *Statistical Quality Control,* was written by Grant and Richard Leavenworth in 1946. In the preface, Grant acknowledged "the particular influence" of Deming's point of view. Deming was engaged in statistical analysis for the United States Census Bureau in the 1940s. This led to his first trip to Japan in 1948 where he was invited to address top business leaders on statistical control. He said that if they followed his advice, they could "take over the world" (Coates 1988).

Another engineer who was active in this field and also conducted courses in the Orient was J. M. Juran. The *Quality Control Handbook* (1951 and 1962), of which Juran was editor-in-chief, contains some of his many writings emphasizing overall quality control. He helped develop the practice of statistical quality control and engineering design principles based on statistical concepts.

LINEAR PROGRAMMING

Economists began to describe economic systems in mathematical terms in the late 1700s. A crude example of a linear programming model was described in Francois Quesnay's *Tableau Economique* where Quesnay attempted to interrelate the roles of the landlord, the peasant, and the artisan. However, linear programming, as it is known today, descended from an input-output method of analysis developed by the economist Wassily Leontief in the 1920s and 1930s. In 1941, Leontief, a professor at Harvard University, interrelated a large number of industries into a single model that permitted parameter estimation in a practical and quantitative way (Leontief 1941). The scope, accuracy, and application area of Leontief-type models were greatly extended by the Bureau of Labor Statistics in the years that followed.

Credit for the development of linear programming belongs to a Soviet mathematician, L. Kantorovich. An English translation of his 1939 paper, "Mathematical Methods of Organizing and Planning Production," appeared in the July 1960 issue of *Management Science*. However, Kantorovich's work was neglected both in his country and in others for many years.

About the same time that Leontief was producing his model, mathmetician John von Neumann was developing concepts that were later published in his paper "A Model of General Economic Equilibrium" published in German in 1936. (The English translation appeared in the "Review of Economic Studies" in 1945.) Von Neumann developed a linear programming model that was more general than Leontief's as it allowed for alternative activities. Later, with Oskar Morgenstern, von Neumann wrote about game theory and economic behavior in *Theory of Games and Economic Behavior* (1944).

One of the first to write about the transportation methods in relation to industrial engineering was Frank Hitchcock. In 1941, his article "The Distribution of a Product from Several Sources to Numerous Localities" appeared in the *Journal of Mathematics and Physics*. The transportation problem has become one of linear programming's most written about techniques because so many methods have been devised for its solution.

Until 1947 linear programming dealt mainly with economic problems. Then, in 1947, it was applied to problems in industry and the military. In that year, George Dantzig developed a general technique for solving linear programming problems that became known as the "simplex method." Problems that could not be solved before could now be solved if enough time was allowed to make the calculations. The development of the computer made the simplex method practical for use in business and industry. Dantzig wrote *Linear Programming and Extensions* (1963) which includes a time table of linear programming's origins and influences. Linear programming has become a technique of a wider field within industrial engineering, operations research.

OPERATIONS RESEARCH

Operations research (OR) had its start in the military. Scientists came into very little contact with business executives until the outbreak of World War II. The military had gone through an evolution similar to that experienced by industry in the Industrial Revolution, but unlike industry, the experience of military leaders was not continuous. New defense equipment and procedures had yet to be used. When the war started, the military was confronted with fighting a war with a system with which they had little real experience. Under these new circumstances, military leaders turned to scientists to help with the operation of the new systems.

In 1937, British scientists were increasingly asked to assist military leaders in learning how to better utilize a new detection device, radar, in locating German aircraft. In solving these new technological problems, a number of scientists were brought together as a team to provide a more general point of view. This type of scientific activity came to be known in Britain as "operational research."

This new activity grew not only in British and U. S. military services, but also in the Canadian and French services. When the war ended European operational research workers moved into government and industrial positions and began to spread the procedures of the new OR techniques. In the United States, military research increased at the war's end, causing OR personnel to remain in the military establishment.

Not until about 1950 did U.S. industry take OR techniques seriously. The revolutionary change brought about by electronic computers helped advance the use of OR techniques.

References

Barnes, R. 1937. *Motion and Time Study*. John Wiley & Sons, Inc. New York.

Barnes, R. 1956. *Work Sampling*. John Wiley & Sons, Inc. New York.

Coates, E. J. 1988. Statistical process control: ignore it at your own risk. *Industrial Management*. July-August. Institute of Industrial Engineers. Norcross, Georgia. 30(4):4.

Dantzig, G. 1963. *Linear Programming and Extensions*. Princeton University Press. Prince ton, New Jersey.

Dodge, H. and H. Romig. 1929. Sampling inspection tables. *Bell System Technical Journal*. October. American Telephone and Telegraph Company. New York.

Fish, J. 1915. *Engineering Economics*. McGraw-Hill, Inc. New York.

Gilbreth, F. and L. 1923. A fourth dimension for measuring skill for obtaining the one best way. *Society of Industrial Engineering Bulletin*. Society of Industrial Engineers.

Goldman, O. 1920. *Financial Engineering*. John Wiley & Sons, Inc. New York.

Grant, E. 1930. *Principles of Engineering Economy*. The Ronald Press. New York.

Grant, E., and R. Leavenworth. 1946. *Statistical Quality Control*. American Society for Quality Control. Milwaukee, Wisconsin.

Grant, E. 1956. *Basic Accounting and Cost Accounting*. McGraw-Hill, Inc. New York.

Hitchcock, F. 1941. The distribution of a product from several to numerous localities. *Journal of Mathematics and Physics*. Massachusetts Institute of Technology. Baltimore, Maryland. 20:224-230.

Holmes, W. 1938. *Applied Time and Motion Study*. The Ronald Press Company. New York.

Industrial Engineering Organization and Practices, Studies in Business Policy, No. 78. National Conference Board, Inc. New York. p. 14.

Industrial Engineering Terminology. 1983. Institute of Industrial Engineers. Norcross, Georgia. p. 168.

Juran, J. 1962. *Quality Control Handbook*. McGraw-Hill, Inc. New York.

Kantorovich, L. 1960. *Management Science*. Institute of Management Science. Baltimore, Maryland.

Leontief, W. 1941. *The Structure of American Economy*, 1919-1929. Harvard University Press. Cambridge, Massachusetts.

Lowry, S., H. B. Maynard, and G. Stegemerten. 1940. *Time and Motion Study*. McGraw-Hill, Inc. New York.

Otis, J. and R. Leukart. 1948. *Job Evaluation*. Prentice-Hall. Englewood Cliffs, New Jersey. National Industrial Conference Board. 1956.

Quesnay, F. 1972. *Quesnay's Tableau Economique*. Edited with translation by Marguerite Kuczynski and Ronald Meek. A.M. Kelley for the Royal Economic Society and the American Economic Association. New York.

Quick, W., R. Koehler, and W. Shea. 1945. *Factory Management and Maintenance*. McGraw-Hill, Inc. New York.

Raymond, F. 1931. *Quality and Economy in Manufacture*. McGraw-Hill, Inc. New York.

Shewhart, W. 1931. *Economic Control of Quality of Manufacturing Product*. Van Nostrand. New York.

Smith, A. 1976. *An Inquiry into the Nature and Causes of the Wealth of Nations*. Edited by R. H. Campbell and A. S. Skinner. Clarendon Press. Oxford. p. 17.

Tompkins, J. 1982. Plant layout. *Handbook of Industrial Engineering*. John Wiley & Sons, Inc. New York.

Von Neumann, J. 1945. A model of general economic equilibrium. *Review of Economic Studies*. London School of Economics and Political Science. Edinburgh. 13(1):1-9.

Weber, A. 1929. *Theory of the Location Industries*. University of Chicago Press. Chicago, Illinois. (Translated by Carl J. Friedrich.)

Wellington, A. 1877. *The Economic Theory of the Location of Railways*. The Railroad Gazette. New York.

Whitin, T. 1953. *The Theory of Inventory Management*. Princeton University Press. Prince ton, New Jersey.

Von Neumann, J. and O. Morgenstern. 1944. *Theory of Games and Economic Behavior*. Princeton University Press. Princeton, New Jersey.

10. Professional Societies and Publications

In 1922, during management week at the annual meeting of the ASME, L. P. Alford presented a paper titled, "Ten Years Progress in Management." In it, he paid tribute to Frederick Taylor as the pioneer in management; to Henry Gantt, who humanized the movement; and to James Dodge, chairman of the board of the Link-Belt Company, as a constructive supporter. All of these men had completed their life work during the same period. Conclusions regarding progress in the management field were mixed, although Taylor's ideas were gaining recognition through scattered application in industry. A major achievement was that good management was depending more on knowledge and fact as a basis for decision making instead of depending on hunches and personal opinion.

The report indicated that during the early 1900s an important step had taken place: the organization of societies promoting scientific management and industrial engineering. Professional societies and their publications provided a forum for new ideas and applications from industry and consultants, and maintained close contact with academic institutions. Their members came from all three of these groups.

THE TAYLOR SOCIETY

The first society in the industrial engineering field was conceived during an informal meeting at the Athletic Club in New York City on November 11, 1910. Frank Gilbreth was the host; Morris Cooke and Robert Kent were among the attendees. Later meetings of this informal organization were usually held monthly at Keenes's Chop House and membership grew to twenty-five or thirty.

Organizations grow out of current events and in 1911 through 1912, public interest in the Eastern Railroad Rate Case hearings influenced the group to form a more solid organization. On November 7, 1912, at the Hotel Astor, the society was named The Society to Promote the Science of Management. Dodge was elected president, Kent became secretary, and Taylor was a charter member.

In 1914, the society began a small journal for the membership which had increased to 110. In 1916, the society was renamed the Taylor

Society in honor of Taylor who had died the year before. On April 1, 1919, the society opened an office in the engineering societies building at 29 West 39th Street in New York City with Harlow Person as managing director.

The society's objective was to promote the scientific study and teaching of the principles governing organized effort through research, discussion, publication, and other appropriate means, for the mutual benefit of the community, labor, management, and the employee.

THE AMERICAN SOCIETY OF SAFETY ENGINEERS

On October 8, 1911, the leading inspectors of insurance companies in New York City were summoned by George Mearns of the Standard Accident Insurance Company to discuss the formation of an association for social purposes. Several inspectors attended and John Anderson of Maryland Casualty Company was elected chairman. The group adopted the name, United Society of Consulting Inspectors (USCI). Meetings were set up on a quarterly basis beginning in November. Membership requirements were established along with an entrance fee of two dollars and quarterly dues of fifty cents.

In 1914, the organization voted to change its name to the American Society of Safety Engineers (ASSE). It was incorporated in 1915 with the objective of promoting the art and science of safety engineering as a profession. In 1919, the society began meeting in the engineering societies building. During this time, considerable interest arose in uniting with the National Safety Council (NSC), and a formal proposal was made by the ASSE in 1920. Negotiations continued, and at the NSC's Louisville Congress in October 1924, the consolidation was completed and the ASSE became known as the American Society of Safety Engineers--Engineering Section, National Safety Council.

In 1947, action was taken to re-establish ASSE as an independent organization. The first election of officers of ASSE as a corporation separate from the NSC was held two years later.

THE SOCIETY OF INDUSTRIAL ENGINEERS

In May 1917, the Western Efficiency Society held a national meeting in Chicago during which the Society of Industrial Engineers was organized. It assisted in the war effort by gathering information from its members about firms capable of manufacturing aircraft parts. Frank, and later Lillian Gilbreth, were members.

At the close of World War I, the society's objectives were changed to those listed in table 7. The membership was divided into six classes:

patrons, professional industrial engineers, professional technical engineers and accountants, managing executives of commercial and industrial activities, engineering educators, and juniors and students.

THE NATIONAL PERSONNEL ASSOCIATION

Taylor emphasized job design and identification which necessitated job specifications and descriptions. This made it possible for qualified workers to be selected to fill vacant positions. This emphasis sparked the beginning of the personnel function in industry.

The first national organization to deal with personnel matters was the National Association of Corporation Schools founded in 1913 to train workers in member firms. By 1917, this organization's scope was broadened to include all of the activities classified as human relations. In 1922, this and other personnel organizations merged to form the National Personnel Association, which in that year had over five hundred individual members and 129 company members.

MANAGEMENT DIVISION OF ASME

In July 1920, a management division was organized within the ASME. This culminated a long history of discussions and proposals dating back to 1886 when Henry Towne presented his paper "The Engineer as Economist." In that paper he developed the concept that the engineer, who uses the dollar symbol in his calculations as much as he uses symbols for feet, minutes, pounds, or gallons, should consider himself an applied economist as well as an engineer. Although, a vast amount of accumulated experience existed in the area of workshop management, it was not available to the world in general (Towne 1886). Towne suggested that a new division of ASME, called the economic section, be organized to originate and promote the dissemination of useful knowledge under headings such as shop management and shop economy. Toward this end, he proposed that a committee be established to consider and make recommendations to the society.

Discussion of this proposal continued until the division was formed in 1920. ASME deserves credit for providing a forum for the presentation of the earliest papers on scientific management. The *Transactions of American Society of Mechanical Engineers* and *The Engineering Magazine* were the early channels of communication and records of developments in the field.

By 1922, there were five societies related to management—four covering engineering and technical aspects and one concerning personnel matters. These societies had a total of over four thousand members as follows:

ASME Management Division	1,740
Society of Industrial Engineers	1,032
The Taylor Society	769
National Personnel Association	500
American Society of Safety Engineers	250

SUBSEQUENT ORGANIZATIONS

The American Management Association grew out the Association of Corporation Schools in 1922. The International Committee of Scientific Management (CIOS) was founded at Geneva, Switzerland in 1924. The American Manufacturers Association was founded in 1929. In 1933, the Association of Consulting Management Engineers was founded. It developed a professional conduct and practice code of standards to offset the poor image that resulted from two decades of charlatanism and pseudo-industrial engineering.

In 1936, the Society for Advancement of Management (SAM) was formed by a merger of the Taylor Society and the Society of Industrial Engineers. It published its own journal, the *Advanced Management Journal*. Many industrial engineers were active members of both SAM and AIIE. (SAM was later acquired by AIIE, becoming the latter's management division.)

Other societies concerned with industrial engineering functions were formed such as the American Society for Quality Control (ASQC) in 1945, and the American Materials Handling Society (AMHS) in 1949.

The American Institute of Industrial Engineers (AIIE) was founded in 1948 in Columbus, Ohio. AIIE, renamed the Institute of Industrial Engineers (IIE) in 1981 to reflect its international membership, remains the official organization representing the industrial engineering profession. It was formed during a period of time when new industrial engineering departments were springing up in universities throughout the country. In 1951, AIIE merged with the Society of Industrial Engineers, a San Francisco Bay area organization of over 150 members headed by Don Malcolm. This was a significant step toward unifying this new profession.

The AIIE's purpose as outlined in its articles of incorporation is shown in table 7 compared with the objectives of the earlier Society of Industrial Engineers founded in 1917. At its founding, membership in the AIIE was limited to those having industrial engineering or related degrees, or students in the process of earning such degrees. Provision was made for affiliates having practical experience in some profession or branch of industry or science which qualified them to cooperate with industrial engineers. More on the founding of AIIE can be found later in this chapter.

TABLE 7.

The Society of Industrial Engineers Objectives, 1917*

1. To furnish a medium for bringing out original contributions to the science of management.
2. To provide an organization through which persons who are applying scientific methods to the solution of the problems of production and distribution may exchange views and coordinate their efforts.
3. To cooperate with other societies.
4. To codify and standardize professional principles and practice.
5. To develop professional standards of the industrial engineer.
6. To promote efficient, energy-conserving management.
7. To enhance the efficiency and prosperity of American industry.

American Institute of Industrial Engineers Purpose, 1948**

1. To maintain the practice of industrial engineering on a professional status.
2. To foster a high degree of integrity among the members of the industrial engineering profession.
3. To encourage and assist education and research in areas of interest to the industrial engineer.
4. To promote the interchange of ideas and information among members of the industrial engineering profession.
5. To serve the public interest by the identification of men qualified to practice as industrial engineers.

*Alford, L. 1922. Ten years progress in management. Transactions of the American Society of Mechanical Engineers. 44:1269.

**From American Institute of Industrial Engineers Articles of Incorporation.

INDUSTRY/PROFESSIONAL PUBLICATIONS

Technical organizations provided the first major publications for industrial engineering. Leading ones were *Transactions of the American Society of Mechanical Engineers* and the Taylor Society's *Bulletin of the Taylor Society*, which continued from 1914 to 1934.

The Engineering Magazine was founded in April 1891 by its first editor, John Dunlap. It included many firsts in the profession. Among these was the first recorded proposal for a new department of industrial engineering made by James Gunn in the article "Cost Keeping: A Subject of Fundamental Importance" in the January 1901 issue. It was the first to list industrial engineering articles by that name; a special index to articles on all phases of industrial engineering was published in its July 1904 issue. In a format intended to be of interest not only to engineers, but to technically minded people as well, it contained reports on engineering projects subsequently published by other engineering societies. Emphasis gradually shifted to factory and manufacturing problems of interest to practical plant men and industrial engineers.

The May 1915 issue of *The Engineering Magazine* included tributes to Taylor, who had died in March. Dunlap added the following note:

> The publication of Mr. Henry R. Towne's appreciative and eloquent tribute to the genius and creative work of Frederick W. Taylor should not pass without immediate mention of the original pioneer and constructive work done by Mr. Towne himself in the field and science of industrial management. Mr. Towne is, in fact, the author of the first proposition that industrial engineering* should be recognized as a science, and that provision should be made for creating and developing a literature relating to it . .
>
> As political economy has been an absorbing study to me since boyhood, when I founded *The Engineering Magazine* in 1891, I determined that industrial economy should be the central feature of the editorial policy; I foresaw plainly that just as we had a science and a whole library of literature upon political economy, so we must now have a recognized science and build up a distinct literature devoted to industrial economy.

*Editor's note: Towne used the expressions *industrial economics* and *shop management*.

The Engineering Magazine Company of New York began publication of *Industrial Management* magazine in 1916. It was written for industrial executives and mechanical engineers and was devoted to labor saving machinery and management. L. P. Alford succeeded Dunlap as chief editor. (Alford also edited *Management Engineering*, a short-lived magazine begun in 1921 to which leading management consultants of the day contributed articles.) In 1928, *Industrial Management* was combined with *Factory* magazine to become *Factory and Industrial Management*.

MAGAZINES TITLED *INDUSTRIAL ENGINEERING*

Over the years, many periodicals have borne the name *Industrial Engineering*, or a derivative thereof. The first magazine bearing the title *Industrial Engineering* appeared in Pittsburgh in 1909. In 1910, it was merged with the *Engineering Digest*, edited by Kent, but kept the name *Industrial Engineering*. (Kent wrote the introduction for Gilbreth's book *Motion Study*, published in 1911. The contents of this book first appeared in serial form in this magazine.)

The *Electrical Review* was acquired by James McGraw's publishing company in 1922 and renamed *Industrial Engineer*. It was edited by Daniel Braymer and dealt with the operation and maintenance of electrical and mechanical systems in mills and factories. The letters "ing" were added to the name in 1928, changing it to *Industrial Engineering*. This magazine was expanded and the name was changed to *Maintenance Engineering* in 1931. (This was later combined with *Factory and Industrial Management* to form *Factory Management and Maintenance* magazine, edited by Robert Rice.)

In January 1949, soon after AIIE was founded, the first issue of the *Journal of Industrial Engineering*, AIIE's official publication, was published with Frank Groseclose as editor-in-chief. (Groseclose has been called the grandfather of industrial engineering in the southeast.) It was published at the Georgia Institute of Technology in Atlanta in the A. French Building. Other early editors were Bob Lehrer and Cecil Johnson.

Such trends in industrial engineering publications indicated the vitality of the concepts that developed from those of the early pioneers and the fact that their hopes were being realized. A comparison of the earliest papers and discussions in the industry, with those in more recent periodicals indicates that instead of a difference in concepts, there is continuity. The newer techniques and facilities available to the modern industrial engineer are expanding the scope of activity rather than changing the original objectives.

RELATED ORGANIZATIONS

The honorary scholastic organization for industrial engineering students is Alpha Pi Mu, founded at the Georgia Institute of Technology. The scholastic society for all engineering students is Tau Beta Pi.

The accrediting agency for industrial engineering and other engineering curricula is the Accreditation Board for Engineering and Technology (ABET), formerly known as the Engineers Council for Professional Development (ECPD). As early as 1912, industrial engineers such as Rautenstrauch, Gilbreth, Emerson, Gantt, and Hathaway were active in ECPD.

After battles for recognition, industrial engineering was accepted by the American Society for Engineering Education (ASEE) and the Engineers Joint Council (EJC). Industrial engineers Rob Roy, Pete Lowman, and Wilson Bentley have served as presidents of ASEE.

THE FOUNDING OF AIIE

The remainder of this chapter highlights a paper by Wyllys Stanton, who served as a professor of industrial engineering at the University of Alabama, the University of Buffalo, and The Ohio State University. He was graduated from Yale University with a bachelor of science degree in 1926, and from Washington University in 1936 with a master of science degree. He was a registered professional engineer in Alabama, Ohio, and New York and a leading force behind the founding of AIIE. He died in 1972.

EARLY HISTORY OF AMERICAN INSTITUTE OF INDUSTRIAL ENGINEERS

By Wyllys Stanton, Founder

The American Institute of Industrial Engineers was formed to provide an organization devoted primarily to the interests and problems of the industrial engineer. The Taylor Society and the Society of Industrial Engineers had been merged into the Society for the Advancement of Management, but it was not devoted to, nor limited to industrial engineers. It included many men whose training and experience could not in any way be described as engineering.

On the evening of January 12, 1948, a group of twelve men met at my home in Columbus, Ohio, in response to an invitation to gather and discuss the problems, methods and potentialities of a new organization specializing in the problems and interests of the industrial engineers. All of this group were either alumni or faculty of Ohio State University.

The points discussed included:

1. Ways an organization of industrial engineers could be useful;
2. Membership requirements;
3. The way organization plans should proceed; and
4. The scope of activities to be attempted.

There seemed to be no question in the founders' minds of the desirability of such an organization. They believed that industrial engineering was an important branch of engineering and just as much in need of an organization devoted to its exclusive representation as civil, mechanical, or electrical engineers. We decided to adopt, subject to possible amendment when a larger group of industrial engineers were brought into the picture, the title American Institute of Industrial Engineers. This choice was based upon the proposition that the organization was to become national in scope at the earliest possible time, and that industrial engineering is a peculiarly American development.

In connection with ways and means of launching the new organization it was considered that two principal avenues lay open: to send out invitations to people who might be interested in helping to form such an organization, or to go ahead and create a going concern, small though it might be, to which all who believed in the idea might join and lend support. Since it was felt that the movement to create the organization was already overdue, and the latter course would be speediest, it was chosen.

The next step was to send letters of invitation to all known industrial engineers in the Columbus area to attend an organization meeting in a classroom at Ohio State University. The response was gratifying. At this second meeting Eldon Raney of Ranco Manufacturing Company was elected as the first president. I was elected executive secretary and a constitution committee was set up. After further discussion of the name it was decided to keep to the original proposal because it was similar to those of the so-called founder societies after which the new group was to pattern its operations. It was also agreed that, because the organization would have little or no service to offer its members in the initial stages, dues should be kept as low as possible.

At a subsequent meeting it was voted to dissolve the group as the national organization and to reform it as the Columbus Chapter of the Institute. Following the original meetings of the Columbus group, extensive mailings were made to a list of industrial engineering alumni of Ohio State and a considerable number of them responded with applications for membership. However, there was the realization of a need for reaching industrial engineers other than the Ohio State group. Therefore, correspondence with the alumni urged them to bring the new organization to the attention of industrial engineers from other schools if they happened to be in contact with any.

From the start it was recognized that the future value and development of the organization depended upon establishing contact with future industrial engineers, while they were still students. Since it had been decided that membership would be limited largely to men who were graduates of industrial engineering curricula that had been accredited by the Engineers' Council for Professional Development, letters were sent to all of the engineering college departments meeting this criterion. Again the response was gratifying, and a number of student chapters were established.

A great advantage of membership in a technical professional society such as the AIIE was the opportunity for association with other members of the profession. Therefore, it was necessary to form chapters rapidly in all centers where there were numbers of industrial engineers and sufficient interest in the project. In those locations where there was an accredited curriculum in industrial engineering, organization of student and senior chapters often proceeded simultaneously, the faculty men forming the nucleus for the senior group.

National Conventions

Recognizing that as membership grew it would be scattered across the country, a constitutional provision held that the AIIE's government should be by the members assembled in national conventions each year. Provision was also made that the executive committee consisting of the national officers should have the power to conduct business between the conventions, so long as they operated within the framework of policies adopted in conventions.

The first annual convention (December 1949) was held in Columbus, Ohio, in the Deshler-Wallick Hotel during the administration of Dwight Gardner, who succeeded Eldon Raney. John Coleman of the Frigidaire Company, who later became national president, served as master of ceremonies. Two awards of the rank of Fellow were made on this occasion, one to John Younger, posthumously, which was received by his widow, and one to Lillian Gilbreth, which was received for her by Elaine Weaver, since Mrs. Gilbreth was not able to be present. The banquet speaker was R. Conrad Cooper, Vice President, Industrial Engineering, of the United States Steel Corporation. Attendance was very good; probably a larger percentage of the total membership than will ever assemble in one room again. The success of this gathering was due to efforts of members of the Columbus Chapter, led by Joe Southern and Stephen Veirs.

The second national convention was held in Cleveland, Ohio, on Saturday, June 16, 1951, at the Statler Hotel. The Cleveland Chapter did a splendid job of making all of the arrangements, including a profitable work measurement conference that preceded the convention. The principal speaker was Ewan Clague, Commissioner of Labor Statistics in the U. S. Department of Labor who talked on comparative labor productivity in the United States and other countries. The rank of Fellow was awarded to Floyd Titler, industrial engineer with the Republic Steel Corporation.

The third national convention was held in Dayton, Ohio, on Friday and Saturday, May 9 and 10, 1952, at the Dayton Biltmore Hotel. The Dayton Chapter also did a splendid job of making all arrangements, and attendance was very good.

The National Headquarters

During the earliest days of the AIIE's development, the operations could be compared fairly with those of a mail order house. Because funds were inadequate to support but very little travel to visit groups of members or prospects, most of the business was necessarily conducted by mail. One of the patterns followed in the operations was that of the American Society for Engineering Education, which had been headquartered on a university campus for many years and was operated with a minimum of clerical help.

For the first three years during which I served as executive secretary, the organization's office was my office at The Ohio State University. During this period, most of the secretarial help was drawn on a part-time basis from the student body. However, this was not entirely satisfactory, because visiting industrial engineers did not always find it possible to go out to the campus, and because the rapid turnover of the student help made continuity of operations difficult.

When I left the university to go into industry in 1950, the office was moved to the Brunson Building at 145 North High Street, in the heart of downtown Columbus. Here, a single small room was obtained for twenty-five dollars per month. At about the same time the office was moved, the organization was fortunate in obtaining the services of Charlotte Patterson, a graduate of Otterbein College who had worked in the administrative offices there. At the time Miss Patterson was working for the Ohio Board of Registration of Professional Engineers and Surveyors and was able to give only part of her time to the AIIE. Later she was able to obtain release from the board and became the AIIE's full-time office secretary. She was responsible for developing many improvements in the headquarters operations.

The space at 145 North High Street soon proved to be inadequate. An opportunity opened up to share a suite of rooms with the Ohio Society of Professional Engineers at 40 West Gay Street. During a part of this joint occupancy I also served as manager of the Ohio Society of Professional Engineer's Engineering Placement Service. Miss Patterson helped in this work, until a new manager and office secretary could be found for that activity.

In September 1952 I left Columbus to take up new duties as a professor of industrial engineering at the University of Alabama. At the Dayton convention in the spring, I had resigned as executive secretary and Joe Southern was elected to succeed me. Much of the summer was devoted to an orderly transfer of the responsibilities.

In December 1952 a fire swept through the building at 40 West Gay Street. Fortunately however, damage to the AIIE office was limited to water-soaked records and out-of-commission furniture and office machines. Miss Patterson and her mother who occasionally helped in the office, took the soaked files home and put them through the electric mangle, thus salvaging almost all of the records. However, it is reported that a faint aroma of smoke still clings to some of the older papers.

Perhaps the most pressing problem was the fact that the organization was homeless, for it was evident that it would take months to restore the building. With the ingenuity and initiative that regularly characterized their actions, Southern, the executive secretary, and Veirs, who had been national treasurer and was serving as assistant executive secretary, set out to search for space. They soon found a site in the Brunson Building, and shortly the organization was re-established in the same building that it had first occupied after leaving the campus.

At this time the loyalty of the membership was demonstrated dramatically. Matt Payne, president, with the assistance of the Cleveland Chapter, got out a letter telling of the fire and and appealing for donations to repair the damage. The response to this appeal was remarkable; enough was collected to fully cover the loss with a small surplus, which was used to help purchase equipment badly needed to cope with the growing volume of work in the headquarters.

WRAP-UP

Many of the firsts, and highlights of the early years of the industrial engineering profession, are listed in the following table.

TABLE 8.
Timetable of Industrial Engineering Events

1881-1882	First experiments in scientific management by Frederick Taylor (time study and metal cutting)
1885	Frank Gilbreth started as a bricklayer's apprentice
1886	Henry Towne's paper, "The Engineer as Economist"
1901	First use of the term "industrial engineering" by James Gunn in *The Engineering Magazine*
1901-1902	First course in shop management by Hugo Diemer at Kansas University.
1903	"Shop Management" by Frederick Taylor published in *Transactions of the American Society of Mechanical Engineers*
1908	First curriculum in industrial engineering by Hugo Diemer at Pennsylvania State College
1910	First meeting of society later named The Society to Promote the Science of Management
	The term "scientific management" used by Louis Brandeis at ICC hearings
	First text in industrial engineering field, *Factory Organization and Administration* written by Hugo Diemer
1911	*Principles of Industrial Engineering* written by Charles Going
	The Principles of Scientific Management written by Frederick Taylor
	Motion Study written by Frank Gilbreth
	American Society of Safety Engineers formed (originally called United Society of Consulting Inspectors)
1912	Testimony by Frederick Taylor before a special committee of the House of Representatives to investigate the Taylor—and other—systems of shop management
1913	First course in scientific management at the University of Tennessee
1914	Gantt Chart applied in the Frankford Arsenal
1915	Frederick Harris' economic lot size equation
	Engineering Economy written by John Fish
	American Society of Safety Engineers incorporated
1917	Society of Industrial Engineers formed
1920	Management division of ASME formed
1922	National Personnel Association formed
	American Management Association formed

1924	Quality control charts by Shewhart
	Motion-time-analysis by A. B. Segur
	International Committee on Scientific Management (CIOS) founded in Geneva, Switzerland
1929	American Manufacturing Association formed
1933	Association of Consulting Engineers formed
1936	Society for the Advancement of Management
1937	Allan Mogensen's Lake Placid work simplification conferences begun
	Operational research activities begun in Britain
1941	*The Structure of American Economy*, 1919-1929 by Wassily Leontief
1946	American Society for Quality Control formed
	Statistical Quality Control by Eugene Grant published
1948	American Institute of Industrial Engineers formed (Wyllys Stanton, founder)
1949	Formation of the American Materials Handling Society

References

Alford, L. P. 1922. Ten years progress in management. *Transactions of the American Society of Mechanical Engineers.* American Society of Mechanical Engineers. New York. 44:1269.

Towne, H. 1886. The engineer as economist. *Transactions of the American Society of Mechanical Engineers.* American Society of Mechanical Engineers. New York. 7:429.

11. New Horizons for Industrial Engineering

The profession of applying science to management was known as scientific management and gradually changed to industrial engineering through the years. With the formation of AIIE in 1948, this name was accepted for the growing number of tools and techniques, which went beyond the orignal concepts of the profession's founders.

Peter Drucker stated in his book, *The Age of Discontinuity*, that the concept of productivity had no meaning until the industrial engineers showed how to measure it. But increasing productivity created the need for the knowledge of how to use it. Those trained in such knowledge involved the new profession of idea men equipped with technical knowledge, rather than brawn or even pragmatic experience, who developed a new type of life and world economy. The most important step toward the "knowledge economy" is described by Drucker in the following quotation that ties the beginning of scientific management to its latest application:

> The most important step toward this 'knowledge economy' was, however, scientific management--that is, the systematic application of analysis and study to manual work, first pioneered by Frederick W. Taylor (1856-1915) in the last decades of the nineteenth century. Taylor, for the first time in history, looked at work itself as deserving the attention of an educated man. Before, work had always been taken for granted, especially by the educated. If they ever thought of it, they knew that work had been ordained--by God or by nature--and that the only way to produce more was to work more and work harder. Taylor saw that this was false. The key to producing more was to 'work smarter.' The key to productivity was knowledge, not sweat.
>
> Taylor did not start out (as most people believe who have never read him) with ideas of efficiency or economy, let alone with the purpose of making a profit for the employer. He started out with a burning social concern, deeply troubled by what he saw as a suicidal conflict between 'labor' and 'capital.' And his greatest impact has also been social. For scientific management (we today would

> probably call it 'systematic work study,' and eliminate thereby a good many misunderstandings the term has caused) has proved to be the most effective idea of this century. It is the only basic American idea that has had worldwide acceptance and impact. Wherever it has been applied, it has raised the productivity and with it the earnings of the manual worker, and especially of the laborer, while greatly reducing his physical efforts and his hours of work. It has probably multiplied the laborer's productivity by a factor of one hundred (Drucker 1968).

Scientific Management - 1947

In the foreword to the 1947 edition of *Scientific Management*, Harlow Person wrote:

> The continuing and, of late, increasing demand for Taylor's papers seems to indicate that not only industrialists but students of social problems sense that they have failed to explore the values of Scientific Management as a great social force. It has been appraised generally in terms of its first emphasis; as a technique for conserving energy and increasing productivity by the use of scientific methods at the individual workplace. But since "Shop Management" was written nearly half a century ago, this technique of conservation has been applied to coordination of all the workplaces of great departments of huge enterprises, and in a few instances to coordination and conservation of the energies of entire enterprises. Some industrialists sense the fact that if we explore what is potential in Scientific Management with larger perspective, we may discover that the philosophy, principles and technique are applicable to conservation problems of entire nations, and perhaps of an entire world (Person 1947).

Recognizing man's problems and his potential, and the role industrial engineering had played so far in society, the following recommendations for the future of industrial engineering were made by Frank Cotton, Jr. when he was president of AIIE:

> 1. We must extend the limits of technological analysis and design to include whatever systems are significant

to the welfare of man. This should include such applications as urban and suburban communities, political and economic systems, and cultural customs and ethical codes.

2. We should utilize considerably more expertise from the social sciences in technology development and integrate the design and application of technology more effectively with the welfare of man and his environment. This should enhance both the effectiveness and efficiency of technology and its service to man.

3. We ought to develop interprofessional groups and functions to focus interdisciplinary expertise on man's major sociotechnical areas of problems and opportunity. Subgroups from professional societies should be formed with the objective of extensive cooperation with other professional groups to develop relevant theory and methods for effective application. Such groups should aggressively seek effective solutions to man's problems and the design of sociotechnical systems that effectively advance man's welfare.

In 1978, the members of AIIE approved a definition of industrial engineering that changed the perspective from individual manufacturing machines or operations to integrated systems of men, materials, energy, and equipment, thus putting a concept of "industrial systems engineering" into prominence.

References

Cotton, F. 1970. An action-oriented concept of man's environment. *Proceedings of the 8th Annual meeting*. Society of Engineering Science. November 9-11, 1970. pp. 667.

Drucker, P. 1968. *The Age of Discontinuity*. Harper and Row Publishers, Inc. New York. p. 271.

Person, H. 1947. Foreword to *Scientific Management*. F. Taylor. Harper and Brothers Publishers. New York. pp. xv-xvi.